Geometry
of
Manifolds

Pure and Applied Mathematics

A Series of Monographs and Textbooks

Edited by

Paul A. Smith and Samuel Eilenberg

Columbia University, New York

Geometry
of Manifolds

By

Richard L. Bishop

Department of Mathematics
University of Illinois
Urbana, Illinois

Richard J. Crittenden

Department of Mathematics
Northwestern University
Evanston, Illinois

1964

ACADEMIC PRESS
NEW YORK AND LONDON

ACADEMIC PRESS INC.
111 Fifth Avenue, New York, New York 10003

United Kingdom Edition published by
ACADEMIC PRESS INC. (LONDON) LTD.
Berkeley Square House, London W.1

LIBRARY OF CONGRESS CATALOG CARD NUMBER: 64-20317

First Printing, 1964
Second Printing, 1967

PRINTED IN THE UNITED STATES OF AMERICA

PREFACE

Our purpose in writing this book is to put material which we found stimulating and interesting as graduate students into book form. It is intended for individual study and for use as a text for graduate level courses such as the one from which this material stems, given by Professor W. Ambrose at MIT in 1958–1959. Previously the material had been organized in roughly the same form by him and Professor I. M. Singer, and they in turn drew upon the work of Ehresmann, Chern, and É. Cartan. Our contributions have been primarily to fill out the material with details, asides and problems, and to alter notation slightly.

We believe that this subject matter, besides being an interesting area for specialization, lends itself especially to a synthesis of several branches of mathematics, and thus should be studied by a wide spectrum of graduate students so as to break away from narrow specialization and see how their own fields are related and applied in other fields. We feel that at least a part of this subject should be of interest not only to those working in geometry but also to those in analysis, topology, algebra, and even probability and astronomy. In order that this book be meaningful, the reader's background should include real variable theory, linear algebra, and point set topology.

To get an idea of the scope of this book we refer to the table of contents and the introductory paragraphs to the chapters. We have not included the study of integration theory, for example, the de Rham's theorems and the Gauss-Bonnet theorem, because we did not wish to get involved in the theory of topological invariants. However, the background for these topics is thoroughly treated, and Morse theory is carried to the point where topology takes over from analysis.

The theorems, lemmas, propositions, and problems are numbered consecutively within each chapter. Our use of these numbers in cross references should be transparent. Thus in the text of Chapter 6, "theorem 7" refers to the seventh theorem in Chapter 6, while "problem 5.4" refers to the fourth problem in Chapter 5. Definitions are generally distinguished only by italics. The word "section" is usually omitted in this usage; that is, an unmodified number reference is to the corresponding section. In this case, the chapter number is always given.

The problems range from trivial to very difficult, from essential to the text to clearly tangential. The subjects of holonomy groups and complex manifolds are developed exclusively in problems. Some problems almost certainly will require recourse to the reference given, namely, problems 1.11, 2.7, 2.13, 2.14, and 8.15.

A brief appendix is provided with a statement of the theorem on existence and uniqueness of solutions of ordinary differential equations most appropriate for our needs.

The reader is referred to [33] and [50] for their extensive bibliographies as well as to their very fine treatment of much of the subject matter of the present text. Italic numbers in brackets are, of course, references to entries in the bibliography.

R.L.B.
R.J.C.

April 1964

CONTENTS

CHAPTER 5

Connexions

CHAPTER 6

Affine Connexions

CHAPTER 7

Riemannian Manifolds

CHAPTER 8

Geodesics and Complete Riemannian Manifolds

CHAPTER 9

Riemannian Curvature

CHAPTER 10

Immersions and the Second Fundamental Form

CHAPTER 11

Second Variation of Arc Length

Geometry
of
Manifolds

CHAPTER 1

Manifolds

In this chapter the basic tools of manifold theory are introduced and the main theorems are stated without proof. Lie derivatives are discussed via local one-parameter groups of transformations, and various interpretations of the bracket of vector fields are given. Frobenius' theorem on the integrability of p-plane distributions is given in outline form [4, 24, 25, 33, 50, 78, 83].

1.1 Introductory Material and Notation

If ϕ is a map of M into N and ψ a map of P into T, then $\psi \circ \phi$ will denote their composition, that is, $\psi \circ \phi$ is ϕ followed by ψ. Here M, N, P, T are any sets and we understand that the domain of $\psi \circ \phi$ is $\phi^{-1}(P) \cap M$ (in particular, $\psi \circ \phi$ may have an empty domain). The same sort of convention, namely, that the domain is the largest meaningful set, will be used in the formation of sums, products, and other combinations of maps. If $U \subset M$ we use $\phi|_U$ for the restriction of ϕ to U.

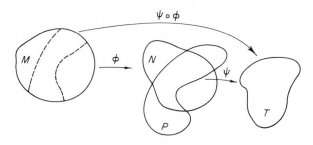

Fig. 1.

The d-dimensional Euclidean space will be denoted by R^d, provided with the usual coordinate functions, $\{u_i\}$, that is, if $t = (t_1, ..., t_d) \in R^d$, then $u_i(t) = t_i$. In the case $d = 1$ we write $R^1 = R$ and $u_1 = u$. If U is open in R^e, then a map $\phi : U \to R^d$ is said to be of *class* C^∞ (written $\phi \in C^\infty$) if the real-valued functions $u_i \circ \phi$, $i = 1, ..., d$, have all kth order continuous partial derivatives for every non-negative k.

C^∞ maps are not necessarily analytic, as is shown by the example: $f(x) = \exp(-1/x^2)$ if $x \neq 0$, $f(0) = 0$. In fact, there exist nontrivial C^∞ real-valued functions on R^d which vanish outside a given compact set. (See [85], pp. 25 and 26 for the construction of C^∞ Urysohn functions.)

Problem 1. Define

$$f(x) = \begin{cases} \exp(-1/x) & \text{if } x > 0 \\ 0 & \text{if } x \leqslant 0. \end{cases}$$

Let $\{r_n\}$ be an ordering of the rational numbers and

$$g(x) = \sum_{n=1}^{\infty} 2^{-n} f(x - r_n).$$

Show that g is C^∞ but nowhere analytic.

1.2 Definition of a Manifold

If X is a Hausdorff topological space, a *d-dimensional coordinate system* in X is a homeomorphism of an open set in X onto an open set in R^d. X is called a *d-dimensional topological manifold* if X is covered by domains of d-dimensional coordinate systems. The domain of a coordinate system ϕ is called the *coordinate neighborhood* and if x is in the coordinate neighborhood of ϕ, ϕ is said to be a *coordinate system at x*.

If ϕ is a coordinate system, we often write $(x_1, ..., x_d)$ for the functions $(u_1 \circ \phi, ..., u_d \circ \phi)$. Either ϕ or $(x_1, ..., x_d)$ will be referred to as a coordinate system.

Let ϕ, ψ be d-dimensional coordinate systems on X. Then ϕ, ψ are *C^∞-related* if $\phi \circ \psi^{-1}$ and $\psi \circ \phi^{-1}$ are of class C^∞. (Fig. 2.)

Consider the following properties of a set of coordinate systems $\mathscr{C}$ on a topological manifold X:

(1) X is covered by the domains of coordinate systems in $\mathscr{C}$.

(2) Every two coordinate systems in $\mathscr{C}$ are C^∞ related.

(3) $\mathscr{C}$ is maximal with respect to (1) and (2).

A C^∞ *manifold* (or just *manifold*) is a pair $(X, \mathscr{C})$, where X is a topological manifold, and $\mathscr{C}$ is a set of coordinate systems satisfying (1),

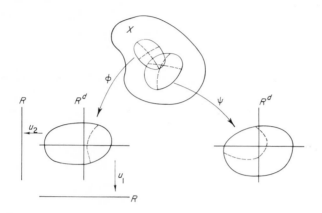

FIG. 2.

(2), and (3). $\mathscr{C}$ is said to be a C^∞ *structure* on X. (We shall usually omit $\mathscr{C}$ in the future, and just write X for a manifold.) A *basis* for the C^∞ structure $\mathscr{C}$ is a subset $\mathscr{C}_0$ of $\mathscr{C}$ satisfying (1) and (2).

Given a set $\mathscr{C}_0$ of coordinate systems on a *set* X satisfying (1) and (2), requiring that they be homeomorphisms defines a topology on X so that X becomes a topological manifold. Then there exists a unique C^∞ structure $\mathscr{C}$ on X with basis $\mathscr{C}_0$ and it is obtained by adjoining all C^∞ related coordinate systems. (We have ignored the assumption that X be Hausdorff.)

If M, N are manifolds, a map $\psi : M \to N$ is *of class* C^∞ ($\psi \in C^\infty$) if for every two coordinate systems ϕ on M, θ on N, the function $\theta \circ \psi \circ \phi^{-1}$ is of class C^∞.

For $\psi : M \to N$ to be of class C^∞ it is sufficient that for every $m \in M$, there are coordinate systems ϕ at m, and θ at $\psi(m)$ such that $\theta \circ \psi \circ \phi^{-1} \in C^\infty$.

If M is paracompact, since partitions of unity which are subordinate to a given covering can be constructed from rational combinations of Urysohn functions, we can get C^∞ partitions of unity. Because C^∞ partitions of unity are an indispensible device for much of the analysis on manifolds, and because Riemannian manifolds, our ultimate object of study, are metrizable, hence paracompact, *we shall assume henceforth that manifolds are paracompact.* It then follows from point-set topology that manifolds are separable and so satisfy the second axiom of countability if connected.

Examples

(1) *Euclidean space.* The ordinary C^∞ structure on R^d is obtained by taking as basis $\mathscr{C}_0 = \{\text{identity}\}$.

(2) *Open submanifolds.* Let M be a manifold with C^∞ structure $\mathscr{C}$, let U be an open subset of M, and let $\mathscr{C}_0 = \{\phi \in \mathscr{C} \mid \text{domain of } \phi \subset U\}$. Then U is a manifold with $\mathscr{C}_0$ as its C^∞ structure. U is called an *open submanifold* of M.

(3) *General linear group.* $Gl(d, R) = \{\text{nonsingular } d \times d \text{ matrices with real entries}\}$ is an open submanifold of R^{d^2}, since

$$Gl(d, R) = R^{d^2} - det^{-1}(0).$$

(4) *Ordinary sphere.* Let $S^d = \{x \in R^{d+1} \mid \Sigma\, u_i^2(x) = 1\}$, and define $\phi : S^d - \{(0, ..., 0, 1)\} \to R^d$, $\psi : S^d - \{(0, ..., 0, -1)\} \to R^d$ by stereographic projection from $(0, ..., 0, 1)$, $(0, ..., 0, -1)$, respec-

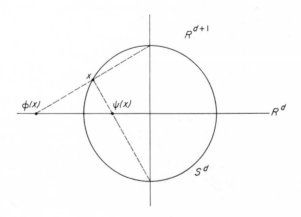

Fig. 3.

tively. Then $\mathscr{C}_0 = \{\phi, \psi\}$ is a basis for a C^∞ structure on S^d. [*Stereographic projection*: $\phi(x)$ is the point where the straight line from $(0, ..., 0, 1)$ through x intersects $u_{d+1}^{-1}(0) = R^d$.]

(5) *Real projective space.* Let P^d be real projective d-space, that is, the collection of straight lines through the origin in R^{d+1}. The natural covering map $\phi : S^d \to P^d$, which takes x into the line through x, induces a C^∞ structure on P^d, that is, there is a unique C^∞ structure on P^d such that ϕ is a C^∞ map with local C^∞ inverses. (See problem 4 below.)

(6) *Low dimensions.* Essentially the only 1-dimensional connected paracompact manifolds are R^1 and S^1. If the restriction of paracompactness is omitted, then this is no longer true, as an example of a long line shows—this is obtained by connecting together an uncountable well-ordered collection of half-closed intervals [*39*, pp. 55 and 56].

A 2-dimensional manifold is called a *surface.* The objects traditionally called "surfaces in 3-space" can be made into manifolds in a standard way. The compact surfaces have been classified as spheres or projective planes with various numbers of handles attached [*80*].

(7) *Product manifolds.* Let M, N be C^∞ manifolds with structures $\mathscr{C}, \mathscr{D}$ and dimensions d, e, respectively. Let $p_1 : M \times N \to M$, $p_2 : M \times N \to N$ be the projections. Then $\mathscr{C}_0 = \{(\phi \circ p_1 , \psi \circ p_2) = (u_1 \circ \phi \circ p_1 , ..., u_d \circ \phi \circ p_1 , u_1 \circ \psi \circ p_2 , ..., u_e \circ \psi \circ p_2) \mid \phi \in \mathscr{C}, \psi \in \mathscr{D}\}$ is a basis for a C^∞-structure on $M \times N$. The same structure could also be obtained by using only a basis for $\mathscr{C}, \mathscr{D}$ in their places. More specific examples are:

Cylinder $= R \times S^1$
2-dimensional (ordinary) torus $= S^1 \times S^1 = T^2$
d-dimensional torus $= S^1 \times \cdots \times S^1$ (d factors) $= T^d$.

Note also that R^{d+e} can be canonically identified with $R^d \times R^e$.

(8) *Non-Hausdorff manifold.* The following example shows that the Hausdorff property of a manifold does not follow from the existence of a C^∞ structure. The underlying point set consists of the interval $(0, 3)$ with topology described in terms of neighborhoods as follows.

The neighborhood of a point in $(0, 1) \cup (1, 2) \cup (2, 3)$ would be as in the topology induced from the reals.

A neighborhood of $i = 1$ or 2 is, for $0 < \epsilon < 1$,

$$(i - \epsilon, i] \cup (2, 2 + \epsilon).$$

We leave as a problem to show that this has a C^∞ structure.

Problem 2. Show that if M is compact then a basis for a C^∞ structure on M must contain more than one coordinate system.

Problem 3. Show that a C^∞ map is necessarily continuous.

A *covering map* $\phi : M \to N$ is a continuous map such that for every $n \in N$ there is a neighborhood U of n such that $\phi^{-1}(U)$ is the disjoint union of neighborhoods of points of $\phi^{-1}(n)$ such that ϕ is a homeomorphism on each such neighborhood [*41*, pp. 89-97; *80*]. ϕ is said to *evenly cover* U and U is said to be a *distinguished neighborhood* of ϕ. When M and N are C^∞ manifolds, then M is said to be a C^∞ *covering* of N if ϕ is a C^∞ map and if the local inverses of ϕ are C^∞ maps.

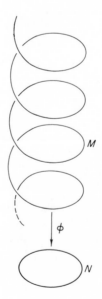

Fig. 4.

Problem 4. Prove that if $\phi : M \to N$ is a covering map and N has a C^∞ structure, then there is a unique C^∞ structure on M such that M is a C^∞ covering of N.

Problem 5. Prove that if $\phi : M \to N$ is a covering map and M has a C^∞ structure such that for every U_i, U_j open sets in M on which ϕ is a homeomorphism and $\phi(U_i) = \phi(U_j)$ we have $(\phi \mid_{U_j})^{-1} \circ \phi$ is a C^∞ map on U_i, then N has a unique C^∞ structure such that M is a C^∞ covering of N.

Problem 6. Prove that if N is a connected C^∞ manifold, then there exists an essentially unique simply connected C^∞ covering of N. (Part of the problem is to show that manifolds satisfy a condition sufficient for the existence of simply connected topological coverings.)

Problem 7. A *d-dimensional complex manifold* is a topological space locally homeomorphic with complex space C^d, and these homeomorphisms are complex analytically related. Make this precise and show that a complex manifold is an even-dimensional (real) manifold.

Examples of complex manifolds are C^d itself, the Riemann surface of a complex analytic function of one complex variable, complex d-dimensional projective space CP^d, and the set of all nonsingular linear transformations of C^d, denoted by $Gl(d, C)$.

1.3 Tangent Space

Let M be a manifold, $m \in M$, and denote by $F(M, m)$ the set of C^∞ real-valued functions with domain a neighborhood of m.

A C^∞ *curve* in M is a map of a closed interval $[a, b]$ into M which can be extended to a C^∞ map of an open interval.

The notion of a tangent can arise from the following considerations. Let γ be a C^∞ curve in M. Then γ gives rise to a linear function $\gamma_*(t) : F(M, \gamma(t)) \to R$ as follows: if $f \in F(M, m)$, $m = \gamma(t)$, then $\gamma_*(t)(f) = (f \circ \gamma)'(t)$, which may be described as a directional derivative of f at m in the direction of γ. [Although $F(M, m)$ is not a

FIG. 5.

linear space, linearity of real-valued functions has an obvious meaning.]
$\gamma_*(t)$ is a derivation, that is, $\gamma_*(t)(fg) = \gamma_*(t)(f)\,g(m) + f(m)\,\gamma_*(t)(g)$.
This linear derivation does everything required of "the tangent to γ"
and subsequently we show that every such linear derivation is
associated with a curve (in fact many curves) in this way.

If $m \in M$, a *tangent to M at m* is a map $t : F(M, m) \to R$ such that

(a) $t(af + bg) = at(f) + bt(g)$
(b) $t(fg) = t(f)\,g(m) + f(m)\,t(g)$, for $a, b \in R$, $f, g \in F(M, m)$.

The tangents at m form a linear space, denoted by M_m. If c is a
constant function, then $t(c) = 0$. We recall that fg and $f + g$ are
the usual product and sum, but defined only on the intersections of
the domains of f and g. Letting 1_U be the function defined only on
U and there equal to 1, we see from (a) and (b) that $t(f1_U) = t(f)$
and hence $t(f) = t(f\,|_U)$; that is, $t(f)$ depends only on the local
behavior of f.

Problem 8.   Prove that for constant function c and neighborhood U
of m that $t(c\,|_U) = 0$. The problem includes proving $t(c) = 0$.

If $\phi = (x_1, ..., x_d)$ is a coordinate system the *partial derivative at
m with respect to x_i*, $D_{x_i}(m)$, is the tangent defined by $(D_{x_i}(m))f =
(\partial(f \circ \phi^{-1})/\partial u_i)(\phi(m))$, which is also denoted by $D_{x_i}f(m)$. When the
coordinates are $\{u_i\}$ on R^d we shall write D_i instead of D_{u_i}.

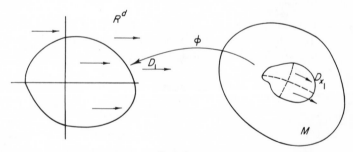

FIG. 6.

It is easily seen that $D_{x_i}x_j(m) = \delta_{ij}$ (Kronecker delta), and, hence,
$\{D_{x_i}(m)\}$ is linearly independent, as can be seen by evaluating a
linear combination on each of the functions x_j in turn.

Problem 9.   Give an example to show that D_{x_i} depends on $x_1, ..., x_d$
and not just on x_i.

Tangents are completely characterized by the following:

Theorem 1. If $(x_1, ..., x_d)$ is a coordinate system at $m \in M$, t a tangent at m, then $t = \Sigma (tx_i) D_{x_i}(m)$.

For the proof, we assume the:

Lemma.† If $f \in F(R^d, a)$, $a = (a_1, ..., a_d)$, then there are functions $g_1, ..., g_d \in F(R^d, a)$ such that $f = f(a) + \Sigma (u_i - a_i) g_i$ in a neighborhood of a [71; 94, p. 221].

Note. From this it follows that $g_i(a) = D_i f(a)$.

Proof of Theorem 1. Let $f \in F(M, m)$, ϕ a coordinate system at m. Then by the lemma, there are g_i such that if $a = \phi(m)$, then $g_i \in F(R^d, a)$, $f \circ \phi^{-1} = f \circ \phi^{-1}(a) + \Sigma (u_i - a_i) g_i$ in a neighborhood of a, and $g_i(a) = D_i(f \circ \phi^{-1})(a)$. Hence, $f = f(m) + \Sigma (x_i - x_i(m)) h_i$ in a neighborhood of m, where $h_i(m) = D_{x_i} f(m)$, $h_i \in F(M, m)$. Therefore,

$$tf = \sum t(x_i) h_i(m) + \sum (x_i(m) - x_i(m)) (th_i)$$

$$= \sum t(x_i) D_{x_i}(m)(f). \qquad \text{QED}$$

Corollary. The dimension of M_m is d, the dimension of M.

(Proof: $\{D_{x_i}(m)\}$ is a basis.)

We have already defined the tangent vector $\gamma_*(t)$ for parameter value t of a C^∞ curve γ in M. We point out that every tangent at m is such a $\gamma_*(t)$. For if $x_1, ..., x_d$ is a coordinate system at m and $s = \Sigma a_i D_{x_i}(m)$, then s is clearly the tangent to the curve given by: $\gamma(t) = $ that point whose coordinates are $x_i(m) + ta_i$.

If $\phi : M \to N$ is C^∞, we define the *differential of ϕ*, $d\phi : M_m \to N_{\phi(m)}$, by: if $t \in M_m$, $f \in F(N, \phi(m))$, then $d\phi(t)(f) = t(f \circ \phi)$.

$d\phi$ is clearly a linear map.

Problem 10. Prove that the following is an alternate definition of $d\phi$: for every C^∞ curve γ in M and parameter value t, $d\phi(\gamma_*(t)) = (\phi \circ \gamma)_*(t)$.

---

† In the case of C^k manifolds (which we have not defined) the corresponding lemma is not true, since the g_i's will not always be C^k. In fact, for C^k manifolds the space of derivations at m is infinite in dimension, so the tangent space is defined to be the space spanned by $\{D_{x_i}(m)\}$ [64A].

Jacobian matrices. If $\phi : M \to N$, $m \in M$, $(x_1, ..., x_d)$ is a coordinate system at m, $(y_1, ..., y_e)$ is a coordinate system at $\phi(m)$, then the matrix of $d\phi$ with respect to the bases $\{D_{x_j}(m)\}$ and $\{D_{y_i}(\phi(m))\}$ is the Jacobian $(D_{x_j}(y_i \circ \phi)(m))$.

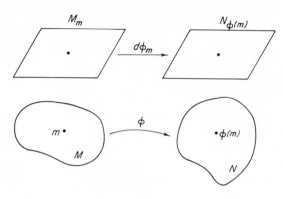

FIG. 7.

Chain rule. If $\phi : M \to N$, $\psi : N \to P$ are C^∞, then $d(\psi \circ \phi) = d\psi \circ d\phi$.

We can rephrase our previous definition of the tangent $\gamma_*(t)$ to a C^∞ curve γ as: $\gamma_*(t) = d\gamma(D(t))$. $(D = D_1 = d/du.)$ Later (3.3) we shall give the collection of all tangents to M, $T(M)$, a C^∞ structure, so that γ_* will become a C^∞ curve in $T(M)$.

Tangent space of a product. If M, N are manifolds, then there is a natural isomorphism τ between $M \times N_{(m.n)}$ and $M_m + N_n$ (direct sum). If $p : M \times N \to M$ and $q : M \times N \to N$ are the projections and $t \in M \times N_{(m,n)}$, then $\tau(t) = dp(t) + dq(t)$.

Now let $\phi : M \times N \to P$ be C^∞. For $(m, n) \in M \times N$, define C^∞ maps

$$\phi_m: N \to P$$
$$\phi^n: M \to P$$

by $\phi_m(n) = \phi^n(m) = \phi(m, n)$. Let $s \in M_m$, $t \in N_n$, and view $s + t$ as an element of $(M \times N)_{(m,n)}$, omitting τ.

Theorem 2. $d\phi(s + t) = d\phi^n(s) + d\phi_m(t)$.

The proof is left as an exercise. (Fig. 8.)

A *diffeomorphism* of M onto N is a one-to-one map $\phi : M \to N$ such that ϕ and ϕ^{-1} are C^∞. Existence of a diffeomorphism is the natural equivalence relation for manifolds. Difficult results of Milnor and others [*44, 52, 53, 58, 84*] show that this equivalence relation is not the same as topological equivalence, at least for manifolds of dimension greater than six. For manifolds of dimensions one, two, and three it is known that both equivalence relations are the same, as is true allegedly also in dimensions four through six according to unpublished work of Cerf.

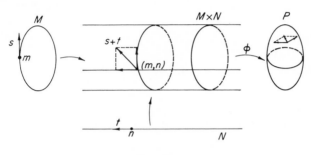

FIG. 8.

It is easy to give examples to show that a C^∞ homeomorphism need not be a diffeomorphism, that is, that the condition that ϕ^{-1} be C^∞ is independent. For any integer $n > 1$ the map $u^{2n-1} : R \to R$ is such an example, since the inverse does not have a derivative at 0. If we change our viewpoint and consider the C^∞ homeomorphism as the identity map on an underlying topological manifold, we get examples of different C^∞ structures on the same space, although the resulting manifolds may be diffeomorphic under another map. The example given then shows that if we take $\{u^{2n-1}\}$ as a basis for a C^∞ structure on R we get different structures for different n's. (However, these manifolds are equivalent in the above sense.)

Characterization of diffeomorphism. $\phi : M \to N$ is a diffeomorphism if and only if ϕ is a C^∞ homeomorphism with range N and for every $m \in M$, ψ a coordinate system at $\phi(m)$, $\psi \circ \phi$ is a coordinate system at m [*25*, p. 75].

Inverse function theorem. Let $(x_1, ..., x_d)$ be a coordinate system at $m \in M$, $f_1, ..., f_d \in F(M, m)$. Then $\phi = (f_1, ..., f_d)$ restricts to a coordinate system at m if and only if $\det(D_{x_j} f_i(m)) \neq 0$, that is, $d\phi$ is nonsingular on M_m [25, p. 70].

Corollary. If $\phi : M \to N$, $\phi \in C^\infty$, $m \in M$ such that $d\phi : M_m \to N_{\phi(m)}$ is one-to-one into, then there is a neighborhood of m in M which is mapped homeomorphically (into) under ϕ. Moreover, if $(y_1, ..., y_e)$ is a coordinate system at $\phi(m)$, then a coordinate system at m may be chosen from restrictions of $y_1 \circ \phi, ..., y_e \circ \phi$ [25, p. 79].

Corollary. If $\phi : M \to N$, $\phi \in C^\infty$, $m \in M$ such that $d\phi : M_m \to N_{\phi(m)}$ is onto, then the image under ϕ of every neighborhood of m in M is a neighborhood of $\phi(m)$ in N. Moreover, if $(y_1, ..., y_e)$ is a coordinate system at $\phi(m)$, then there are C^∞ functions $x_{e+1}, ..., x_d$ defined in a neighborhood of m such that $(y_1 \circ \phi, ..., y_e \circ \phi, x_{e+1}, ..., x_d)$ is a coordinate system at m [25, p. 80].

Corollary. (*Inverse function theorem for manifolds.*) If $\phi : M \to N$, $\phi \in C^\infty$, $m \in M$, then ϕ is a diffeomorphism of an open neighborhood of m onto an open neighborhood of $\phi(m)$ if and only if $d\phi$ is an isomorphism onto at m [25, p. 80].

Corollary. If $\phi : M \to N$, $\phi \in C^\infty$, $d\phi = 0$ everywhere, and M is connected, then ϕ is constant [25, p. 80].

Problem 11. Measure zero is a sensible notion on manifolds: $S \subset M$ has *measure zero* if for every coordinate map ϕ, $\phi(S) \subset R^d$ has measure zero. Prove *Sard's theorem*:

If $\psi : N \to M$ is C^∞, then $S = \{ m \in M \mid m = \psi(n)$ for some n such that $d\psi : N_n \to M_m$ is not onto$\}$ has measure zero [79].

Problem 12. Prove the following generalization of the first corollary above.

Let C be a compact subset of M, $\phi : M \to N$, $\phi \in C^\infty$, such that ϕ is one-to-one on C and for every $m \in C$, $d\phi_m$ is one-to-one into. Then there is a neighborhood of C which is mapped homeomorphically under ϕ [83, p. VI-45].

Differentials of functions. Every element f of $F(M, m)$ gives rise, via its differential, to an element of the dual space $M_m{}^*$ of M_m as follows: we may identify $R_{f(m)}$ with R [$aD(f(m))$ is identified with a], and hence $df : M_m \to R_{f(m)} \simeq R$. Note that under this identification, if

$t \in M_m$, then $df(t) = t(f)$. Now if $(x_1, ..., x_d)$ is a coordinate system at m, then we have $dx_i(D_{x_j}(m)) = \delta_{ij}$, so $(dx_1, ..., dx_d)$ forms a basis of $M_m{}^*$ dual to the $D_{x_j}(m)$. In fact, for any $f \in F(M, m)$, $df = \Sigma\, D_{x_i} f(m)\, dx_i$.

1.4 Vector Fields

A *vector field*, X, is a function defined on a subset E of a manifold M which assigns at each point $m \in E$ an element $X(m)$ of M_m.

If X is a vector field, $f \in F(M, m)$, then Xf is the function defined on the intersection of the domains of X and f by: $Xf(n) = X(n)(f)$.

A vector field X is *of class* C^∞ if its domain is open and for every m in the domain of X and $f \in F(M, m)$, $Xf \in F(M, m)$ also.

Frequently we shall consider a C^∞ vector field X to be the map on C^∞ functions given by $f \to Xf$, since X is entirely determined this way by varying f.

If $(x_1, ..., x_d)$ is a coordinate system, then D_{x_i} is a C^∞ vector field. If X is a vector field with its domain contained in the coordinate system, we may write $X = \Sigma f_i D_{x_i}$, where the f_i are real-valued functions.

Problem 13. Prove that $X \in C^\infty$ if and only if the $f_i \in C^\infty$.

If f is a C^∞ map of M into R^e, so $f = (f_1, ..., f_e)$ with f_i real valued, and X a vector field on M, we write Xf for $(Xf_1, ..., Xf_e)$. Similarly we define $tf \in R^e$, for a tangent t.

It is clear that if $X \in C^\infty$, then $Xf \in C^\infty$.

If X, Y are C^∞ vector fields, then we define a C^∞ vector field $[X, Y]$, called the *bracket of X and Y*, on the intersection of their domains by $[X, Y] = XY - YX$. Multiplication of vector fields here is composition of their action on functions.

Problem 14. If X and Y are C^∞ vector fields prove:

(a) $[X, Y]$ actually is a vector field.
(b) XY is not a vector field unless one of them is 0.
(c) If f and g are real-valued C^∞ functions, then

$$[fX, gY] = fg[X, Y] + f(Xg)\,Y - g(Yf)\,X.$$

Coordinate expression for bracket. If X, Y are C^∞ vector fields, $(x_1, ..., x_d)$ a coordinate system, $X = \Sigma f_i D_{x_i}$, $Y = \Sigma g_i D_{x_i}$ on the common part of the domains, then

$$[X, Y] = \sum_{i,j} (f_i D_{x_i} g_j - g_i D_{x_i} f_j) \, D_{x_j}.$$

The bracket operation is bilinear with respect to real coefficients. It is also skew-symmetric, that is, $[X, X] = 0$, or equivalently, $[X, Y] = -[Y, X]$.

Jacobi identity. If X, Y, Z are C^∞ vector fields, then

$$[[X, Y], Z] + [[Y, Z], X] + [[Z, X], Y] = 0.$$

Another way of expressing this is to say that the map $Y \to [X, Y]$, the *Lie derivative with respect to* X, is a derivation of the algebra of C^∞ vector fields, where multiplication in that algebra is bracket: $[X, [Y, Z]] = [[X, Y], Z] + [Y, [X, Z]]$.

Vector fields and maps. If $\phi : M \to N$, $\phi \in C^\infty$, X, Y vector fields on M, N, respectively, then X, Y are ϕ *related* if for every m in the domain of X, $d\phi(X(m)) = Y(\phi(m))$.

If ϕ has the property that $d\phi$ is one-to-one at every point then ϕ is called *regular*.

Problem 15. If ϕ is regular and Y is a C^∞ vector field on N such that for every $m \in \phi^{-1}$(domain of Y), $Y(\phi(m)) \in d\phi(M_m)$, then there is a unique C^∞ vector field X on M which is ϕ related to Y [25, p. 84].

Note that in general, if $\phi : M \to N$, $\phi \in C^\infty$, X a vector field on M, then $d\phi \, X$ is not defined. Namely, if $m, n \in M$ are such that $\phi(m) = \phi(n)$ but $d\phi(X(m)) \neq d\phi(X(n))$, then $d\phi \, X$ is not single valued at $\phi(m)$.

Problem 16. Give an example for which $d\phi \, X$ is not defined when $M = R$, $N = S^1$, and ϕ is regular.

Problem 17. *Brackets and maps.* If $\phi : M \to N$, $\phi \in C^\infty$, X_1, X_2 C^∞ vector fields on M, Y_1, Y_2 C^∞ vector fields on N such that X_i is ϕ related to Y_i ($i = 1, 2$), then $[X_1, X_2]$ is ϕ related to $[Y_1, Y_2]$ [25, p. 85].

Integral curves. If X is a C^∞ vector field, then γ is *the integral curve of X starting at m* if $\gamma(0) = m$ and for every t in the domain of γ, $\gamma_*(t) = X(\gamma(t))$. The existence of integral curves and their essential uniqueness are immediate consequences of the corresponding theorems for solutions of systems of ordinary differential equations: at any point we merely take a coordinate system and transfer everything to an open set of R^d. By *essential uniqueness* we mean: if γ and τ are integral curves starting at m, then their restrictions to the common interval of definition are the same.

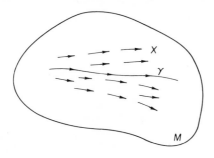

FIG. 9.

Problem 18. Let $(x_1, \ldots, x_d)$ be a coordinate system at m, γ the integral curve of X starting at m, $f_i = x_i \circ \gamma$, so that each f_i is a real-valued C^∞ function defined on an interval of the real axis containing 0. Show that the equation $\gamma_*(t) = X(\gamma(t))$ is equivalent to an equation involving the f_i and their derivatives, so that the problem of finding integral curves is equivalent to solving systems of ordinary differential equations as claimed.

Local one-parameter group. Let X be a C^∞ vector field. We associate with X a *local one-parameter group of transformations* T which for every $m \in M$ and real number t sufficiently close to 0 assigns the point $T(m, t) = \gamma(t)$, where γ is the integral curve of X starting at m. By theorems in differential equations on the dependence of solutions on initial conditions, for every m there is a positive number c and a neighborhood U of m such that T is defined and C^∞ on $U \times (-c, c)$. Since the real numbers used as the second variable of T are parameter values along a curve, they satisfy an additive property: if $n \in U$, $t, s, s + t \in (-c, c)$, then $T(T(n, t), s) = T(n, s + t)$.

Conversely, if we are given a C^∞ map having domain of the same type as T and satisfying the additive property, then again calling it T, we get a vector field having T as its local one-parameter group as follows: let the injection j_m be defined by $j_m(t) = (m, t)$. Then at m the value of the vector field is $X(m) = (T \circ j_m)_*(0)$.

Problem 19. Let $X = u_1 D_1 + u_2 D_2$. Find explicit equations for $T : R^2 \times R \to R^2$. Do the same for $Y = -u_2 D_1 + u_1 D_2$.

Lie derivatives. Functions, differentials, vector fields, and other geometric objects on M are acted upon by transformations: functions by composition with the transformation, differentials by composition with the differential of the transformation, and vector fields by the

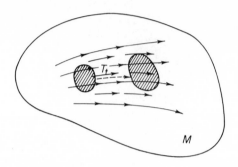

FIG. 10.

action of the differential of the transformation. It should be noticed that functions and differentials are *pulled back* from the range to the domain of the transformation whereas vector fields are *pushed forward* from the domain to the range. Now if we consider the values of one of these geometric objects along γ, the integral curve of X starting at m, then by using the transformations $T_t = T \circ {}'j_t$, where ${}'j_t(n) = (n, t)$, we can get a curve of values at m, that is either a real-valued function of t, or a curve in $M_m{}^*$ or in M_m. Since in each case the values are in a vector space we can differentiate. The derivative at 0 is then called the *Lie derivative of the object with respect to X at m.*

In the case of a function f, since we are just differentiating the value of f along γ with respect to the parameter of γ, we get the tangential derivative with respect to the curve, that is, we get $X(m)f$.

In the case of a differential of a function, df, the curve in $M_m{}^*$ is given by $t \to df(\gamma(t)) \circ (dT_t)_m$, where $(dT_t)_m$ denotes the restriction of dT_t to M_m. It is not difficult to show that the Lie derivative in this case is $(d(Xf))_m$ [*24*, p. 75].

In the case of a vector field Y we must pull back the value at $\gamma(t)$ to m, which we do via dT_{-t}, so the curve in M_m is $t \to dT_{-t}(Y(\gamma(t)))$.

Since we shall return to this case in the chapter on bundles we state the result as a theorem; the result agrees with the terminology already introduced above under Jacobi identity.

Theorem 3. The Lie derivative of Y with respect to X is $[X, Y]$.

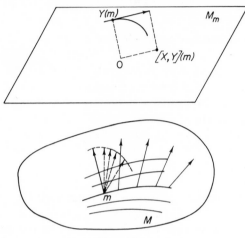

FIG. 11.

Proof. We must show that for $f \in F(M, m)$

$$\frac{d}{dt}(0) \{dT_{-t}(Y(\gamma(t)))f\} = X(m)\, Yf - Y(m)\, Xf.$$

Noticing that $dT_{-t}(Y(\gamma(t)))f = Y(\gamma(t))(f \circ T_{-t})$, we see that we must consider derivatives with respect to Y, and hence we introduce S, the one-parameter group associated with Y. Then if we define $G : V \rightarrow R$, where V is a neighborhood of $(0, 0)$ in R^2, by

$$G(t, r) = f(T(S(\gamma(t), r), -t))$$

it is immediate from the definition of S that

$$Y(\gamma(t))(f \circ T_{-t}) = D_2 G(t, 0),$$

and hence, that

$$\frac{d}{dt}(0)\, dT_{-t}(Y(\gamma(t)))f = D_1 D_2 G(0, 0).$$

Now letting $H(t, r, s) = f(T(S(\gamma(t), r), s))$, it follows from the chain rule that $D_1 D_2 G(0, 0) = D_1 D_2 H(0, 0, 0) - D_2 D_3 H(0, 0, 0)$. Since $H(t, r, 0) = f(T(S(\gamma(t), r), 0)) = f(S(\gamma(t), r))$, it follows that $D_2 H(t, 0, 0) = Yf(\gamma(t))$, and thus $D_1 D_2 H(0, 0, 0) = X(m) Yf$. Since $H(0, r, s) = f(T(S(m, r), s))$, it follows that $D_3 H(0, r, 0) = Xf(S(m, r))$, and thus $D_2 D_3 H(0, 0, 0) = Y(m) Xf$. QED

Theorem 4. *Geometrical interpretation of bracket.* Let X, Y be C^∞ vector fields both defined at $m \in M$. We shall define a curve which has $[X, Y](m)$ as a limit of its tangents. Let g_1 be the integral curve of X starting at m. Then for sufficiently small positive c the remainder of the construction works. Let g_2 be the integral curve of Y starting at $g_1(c)$. Let g_3 be the integral curve of $-X$ starting at $g_2(c)$, g_4 the integral curve of $-Y$ starting at $g_3(c)$. Define a curve g by $g(c^2) = g_4(c)$. Then $[X, Y](m) = \lim_{t \to 0+} g_*(t)$, that is, for every $f \in F(M, m)$, $[X, Y](m) f = \lim_{t \to 0+} g_*(t) f$.

Proof. Define maps h_1, h_2, h_3 of a neighborhood of $(0, 0)$ in R^2 into M by: $h_1(t, c) = g_2(t)$, $h_2(t, c) = g_3(t)$, $h_3(t, c) = g_4(t)$. These maps are C^∞ because they can be expressed as compositions of the one-parameter groups of X and Y; they make explicit the dependence of g_2, g_3, g_4 on c. Also let $h(t) = h_3(t, t)$, so that $g(t^2) = h(t)$.

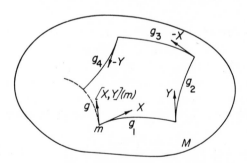

FIG. 12.

First it will be shown that $h_*(0) = 0$, and then according to problem 20, which follows, it will remain to prove that

$$2[X, Y]f(m) = (f \circ h)''(0).$$

The following facts are immediate from the definitions:

(a) $D_2(f \circ h_1)(0, t) = Xf(h_1(0, t))$ (b) $D_1(f \circ h_1) = Yf \circ h_1$

(c) $D_1(f \circ h_2) = -Xf \circ h_2$ (d) $D_1(f \circ h_3) = -Yf \circ h_3$

(e) $h_3(0, t) = h_2(t, t)$ (f) $h_2(0, t) = h_1(t, t)$.

Then we have

$$(f \circ h)'(0) = D_1(f \circ h_3)(0, 0) + D_2(f \circ h_3)(0, 0) \qquad \text{[chain rule]}$$

$$= -Yf(m) + D_1(f \circ h_2)(0, 0) + D_2(f \circ h_2)(0, 0) \qquad \text{[(d), (e)]}$$

$$= -Yf(m) - Xf(m) + D_1(f \circ h_1)(0, 0) + D_2(f \circ h_1)(0, 0)$$
$$\text{[(c), (f)]}$$

$$= 0 \qquad \text{[(a), (b)]}$$

$$(f \circ h)''(0) = D_1{}^2(f \circ h_3)(O) + 2D_2 D_1(f \circ h_3)(O) + D_2{}^2(f \circ h_3)(O)$$
$$\text{[}(O) = (0, 0)\text{]}$$

$$= Y^2 f(m) - 2D_1(Yf \circ h_2)(O) - 2D_2(Yf \circ h_2)(O)$$
$$+ D_1{}^2(f \circ h_2)(O) + 2D_2 D_1(f \circ h_2)(O) + D_2{}^2(f \circ h_2)(O)$$

$$= Y^2 f(m) + 2XYf(m) - 2D_1(Yf \circ h_1)(O) - 2D_2(Yf \circ h_1)(O)$$
$$+ X^2 f(m) - 2D_1(Xf \circ h_1)(O) - 2D_2(Xf \circ h_1)(O)$$
$$+ D_1{}^2(f \circ h_1)(O) + 2D_2 D_1(f \circ h_1)(O) + D_2{}^2(f \circ h_1)(O)$$

$$= Y^2 f(m) + 2XYf(m) - 2Y^2 f(m) - 2XYf(m)$$
$$+ X^2 f(m) - 2YXf(m) - 2X^2 f(m)$$
$$+ Y^2 f(m) + 2XYf(m) + X^2 f(m)$$

$$= 2XYf(m) - 2YXf(m). \qquad \text{QED}$$

Problem 20. Let h be a C^∞ curve such that $h_*(0) = 0$, and define g on small positive numbers by $g(t^2) = h(t)$, and let $m = h(0)$. Show that:

(1) The map defined on $F(M, m)$ by $f \to (f \circ h)''(0)$ is a tangent at m.

(2) The same tangent is $2 \lim_{t \to 0+} g_*(t)$, where the meaning of lim is the same as above.

If we call this the 2nd order tangent to h at a point where the 1st order tangent vanishes, generalize (1) so as to define an nth order tangent to h at a point where the 1st order, 2nd order, ..., $(n-1)$st order tangents vanish, and prove a result analogous to (2).

Problem 21. If X is a C^∞ vector field and $X(m) = 0$, show that the integral curve of X starting at m is the constant curve: $\gamma(t) = m$ for every t. Hence if γ is a C^∞ curve with $\gamma_*'(0) = 0$, then γ_* does not have an extension to a C^∞ vector field unless γ is a constant curve.

Problem 22. For X as in problem 19 and $f = u_2$ show explicitly that the Lie derivative of df with respect to X is $d(Xf) = df$.

Problem 23. Use theorem 3 and also theorem 4 to show that $[X, Y] = 0$, where X and Y are as in problem 19.

It is a consequence of the following theorem that if $[X, Y] = 0$ in a neighborhood of m then the broken curves of theorem 4 are actually closed for sufficiently small t, that is, $g(t) = m$ for all t near 0.

Theorem 5. Let X_i, $i = 1, ..., e$, be C^∞ vector fields defined and linearly independent in a neighborhood of $m \in M$, where M is a d-dimensional manifold, and such that $[X_i, X_j] = 0$ in that neighborhood for every i, j. Then there is a coordinate system $(x_1, ..., x_d)$ such that X_i coincides with D_{x_i} in the coordinate neighborhood $(i \leqslant e)$ and $x_i(m) = 0$, $i = 1, ..., d$.

Proof. Choose a coordinate system $(y_1, ..., y_d)$ such that $y_i(m) = 0$, all i, and such that $X_1(m), ..., X_e(m), D_{y_{e+1}}(m), ..., D_{y_d}(m)$ are a basis of M_m. Define $\psi : U \to M$, where U is a neighborhood of $0 \in R^d$ by the conditions

$$y_i(\psi(0, 0, ..., 0, a_{e+1}, ..., a_d)) = \begin{cases} a_i, & i > e \\ 0, & i \leqslant e. \end{cases}$$

$t \to \psi(0, 0, ..., 0, t, a_{i+1}, ..., a_d)$ is the integral curve of X_i starting at $\psi(0, 0, ..., 0, a_{i+1}, ..., a_d)$, $i \leqslant e$.

The map ψ can also be expressed as compositions of the one-parameter groups of $X_1, X_2, ..., X_e$ and of the inverse of the y-coordinate map, so for this reason or directly from theorems on differential equations, we see that ψ is C^∞.

From the definition it is immediate that $d\psi(D_i(0)) = X_i(m)$, $i \leqslant e$, and $d\psi(D_i(0)) = D_{y_i}(m)$, $i > e$. By the inverse function theorem ψ^{-1} is defined and C^∞ in a neighborhood of m and thus defines a coordinate system $(x_1, ..., x_d)$ such that $x_i(m) = 0$, all i.

Because the integral curves of D_1 always correspond to integral curves of X_1 under ψ, it is clear that $D_{x_1} = X_1$ in the coordinate neighborhood. We complete the proof by showing by induction on i

that $X_i x_j = \delta_{ij}$. Since the integral curves of D_i starting at points of the form $(0, ..., 0, a_i, ..., a_d)$ correspond to integral curves of X_i under ψ, $X_i x_j = \delta_{ij}$ at points of the form $\psi(0, ..., 0, a_i, ..., a_d)$. But $X_k X_i = X_i X_k$, so that for $k < i$, $X_k(X_i x_j) = X_i(X_k x_j) = 0$ by the induction assumption. Thus $X_i x_j$ is constant along the integral curves of X_k, $k < i$. But every point in the coordinate neighborhood can be reached via chains of such integral curves starting at points of the form $\psi(0, ..., 0, a_i, ..., a_d)$, so $X_i x_j = \delta_{ij}$ everywhere.   QED

Problem 24.   Find the expression for ψ in the above proof as compositions of the one-parameter groups of X_1, X_2, ..., X_e and of the inverse of the y-coordinate map.

1.5 Submanifolds

Let M be a C^∞ manifold. A manifold N is a *submanifold* of M if there is a one-to-one C^∞ map $i : N \to M$ such that di is one-to-one at every point. We call i an *imbedding* and say that N is *imbedded* in M by i.

Frequently we start with a subset of M, describe a C^∞ structure for it, so that it becomes a submanifold under the inclusion map. For example, in this way an open submanifold of M becomes a submanifold of M. The d-sphere as a submanifold of R^{d+1} and R^d, imbedded as $R^d \times \{0\}$ in R^{d+e}, are examples where the dimension of the submanifold is less than that of the containing manifold.

Problem 25.   Compute that the inverse of stereographic projection $R^d \to R^{d+1}$ has one-to-one differential everywhere, and point out how this shows S^d is a submanifold of R^{d+1}.

Striped pants example.   This example illustrates that the map $i : N \to M$, although continuous, need not be a homeomorphism into, that is, the topology on $i(N)$ induced as a subset of M need not be the same as the manifold topology of N. Let M be the two-dimensional torus, $N = R$, imbedded as a dense one-parameter subgroup of M [that is, for $x \in R$, $i(x) = (\exp icx, \exp idx) \in S^1 \times S^1$, where c/d is irrational]. Thus N is a submanifold of M, but $i(N)$ is not even closed in M, and hence not locally compact.

By altering the C^∞ structure of an open submanifold we can get examples of subspaces (topological) which are not submanifolds, since the inclusion map need not be C^∞, or, if it is C^∞ it may not be regular at some points.

Problem 26. Let $\psi : N \to M$ be C^∞ with $\dim M < \dim N$, and let S be the set of measure zero defined in problem 11. Prove that if $m \notin S$ then $P = \psi^{-1}(m)$ with the topology induced from N has a unique C^∞ structure so that it becomes a submanifold of N under the inclusion map [*88*].

Problem 27. For $\psi = u_2{}^3 + u_1{}^3 - 3u_1u_2 : R^2 \to R$ find what S is and show why $\psi^{-1}(m)$ is not a submanifold in the cases where $m \in S$.

When ψ is a real-valued function on R^3 the submanifolds obtained as in problem 26 are the traditional surfaces in 3-space.

Problem 28. Let $f : P \to M$ be C^∞, $i : N \to M$ a submanifold, such that $f(P) \subset i(N)$. Let $g = i^{-1} \circ f : P \to N$.

(a) If g is continuous show that g is C^∞.

(b) Find an example for which g is not continuous.

Restrictions of functions. If $i : N \to M$, N a submanifold of M, $f \in F(M, m)$, $m \in i(N)$, then $f \circ i \in F(N, i^{-1}(m))$ is the *restriction of f to N*. Conversely, if $g \in F(N, n)$, then there is a neighborhood U of n in N and $f \in F(M, i(n))$ such that $f \circ i\,|_U = g\,|_U$. However, this is not true globally. For example, in the striped pants example the identity function $u : R \to R$ is not the restriction of any continuous real-valued function on the torus M.

Restrictions of vector fields. If $i : N \to M$, N a submanifold of M, X a C^∞ vector field on M such that for all $n \in N$, $X(i(n)) \in di(N_n)$, then by problem 15 there is a C^∞ vector field Y on N such that $di(Y(n)) = X(i(n))$. *Y is the restriction of X to N.* Note that by problem 17 the bracket of restrictions is the restriction of the bracket.

Vector fields on N can be extended locally, that is, locally realized as restrictions, in the same way as functions. In particular, if γ is a C^∞ curve in M with nonvanishing tangent γ_*, then locally γ_* can be extended to a vector field in M (cf. problem 21).

1.6 Distributions and Integrability

A *p-dimensional distribution* on a manifold $M(p \leqslant \dim(M))$ is a function θ defined on M which assigns to each $m \in M$ a p-dimensional linear subspace $\theta(m)$ of M_m. A p-dimensional distribution θ on M

is of *class* C^∞ at $m \in M$ if there are C^∞ vector fields $X_1, ..., X_p$ defined in a neighborhood U of m and such that for every $n \in U$, $X_1(n), ..., X_p(n)$ span $\theta(n)$. An *integral manifold* N of θ is a submanifold of M such that $di(N_n) = \theta(i(n))$ for every $n \in N$. We say that a vector field X *belongs* to the distribution θ and write $X \in \theta$, if for every m in the domain of X, $X(m) \in \theta(m)$. A distribution θ is *involutive* if for all C^∞ vector fields X, Y which belong to θ, we have $[X, Y] \in \theta$. A distribution θ is *integrable* if for every $m \in M$ there is an integral manifold of θ containing m. We sometimes write θ_m for $\theta(m)$.

That an integrable C^∞ distribution is involutive is easily seen by problem 17 or by theorem 4. The latter also gives us an insight into the reason the converse is true. Going around a one-parameter family of rectangles having sides tangent to a distribution might give a curve which has a tangent not in the distribution, but in the case of an involutive distribution this cannot happen; thus when the distribution is involutive integral manifolds can be generated by going along the various integral curves of vector fields belonging to the distribution.

Every one-dimensional C^∞ distribution is both involutive and integrable, by the existence of integral curves.

Problem 29.   Let P, Q, R be C^∞ functions on an open set U in R^3 which do not all vanish simultaneously. Let $\theta(m)$ be the linear subspace of R_m^3 which is orthogonal to $(P(m), Q(m), R(m))$. Show that θ is involutive if and only if

$$P(D_3Q - D_2R) + Q(D_1R - D_3P) + R(D_2P - D_1Q) = 0.$$

An integral manifold is a solution of the equation

$$P \, du_1 + Q \, du_2 + R \, du_3 = 0.$$

The following theorem, the converse mentioned above, is proved most easily by first getting the local form. The form stated here then follows by taking unions. The dual formulation in terms of differential forms is the classical theorem of Frobenius.

Theorem 6.   A C^∞ involutive distribution θ on M is integrable. Furthermore, through every $m \in M$ there passes a unique maximal connected integral manifold of θ and every other connected integral manifold containing m is an open submanifold of this maximal one [*25*, p. 94].

The local theorem gives more information as to how the integral manifolds are situated with respect to each other:

Theorem 7. If θ is a C^{∞} involutive distribution on M, $m \in M$, then there is a coordinate system $(x_1, ..., x_d)$ on a neighborhood of m, such that $x_i(m) = 0$ and for every m' in the coordinate neighborhood the *slice* $\{p \in M \mid x_i(p) = x_i(m')$ for every $i > e\}$ $(e = \dim \theta)$ is an integral manifold of θ, when given the obvious manifold structure induced by the coordinate map.

Outline of proof. It suffices to construct a coordinate system such that for every X belonging to θ, $Xx_i = 0$ for $i > e$.

Let $Y_1, ..., Y_e$ be C^{∞} vector fields which span θ in a neighborhood of m. Choose coordinates y_i so that $Y_e = D_{y_1}$. (This can be done by theorem 5 applied to one vector field.) Let $Y_i' = Y_i - (Y_i y_1) Y_e$, $i < e$, $Y_e' = Y_e$. Then since $Y_i' y_1 = 0$, $i < e$, it follows that $[Y_i', Y_j'] y_1 = 0$ for all i, j, and then that $[Y_i', Y_j']$ can be written as a linear combination (with real-valued C^{∞} functions as coefficients) of $Y_1', ..., Y_{e-1}'$. In particular, Y_i', $i < e$, span an involutive $(e - 1)$-dimensional distribution. Repeating this process $e - 2$ more times gives vector fields $X_1, ..., X_e$ which span θ and such that the first i, $i \leqslant e$, have brackets expressible as a linear combination of the first $i - 1$ ($X_e = Y_e$, $X_{e-1} = Y_{e-1}'$, etc.).

Using the X_i just constructed we follow, step by step, the proof of theorem 5 except in the last paragraph, where we want to show less ($X_i x_j = 0$ only for $j > e$), and hence need less ($X_k X_i = X_i X_k$ + linear combination of the X_h with $h < i$).

Problem 30. Let $\theta_1, ..., \theta_h$ be complementary C^{∞} distributions on M, that is, M_m is the direct sum of the $\theta_i(m)$ for every $m \in M$. Prove that if all the θ_i are integrable, then they are simultaneously integrable in the following sense: at every $m \in M$ there is a coordinate system $(x_1, ..., x_d)$ such that $D_{x_1}, ..., D_{x_{d_1}}$ span $\theta_1, ..., D_{x_{d_{h-1}+1}}, ..., D_{x_d}$ span θ_h, where $d_i - d_{i-1} = \dim \theta_i$ ($d_0 = 0$, $d_h = d$).

Problem 31. Prove the following addition to problem 28:

(c) If N is an integral manifold of a distribution on M, then g is continuous.

CHAPTER 2

Lie Groups

The basic definitions and theorems on Lie groups and Lie algebras are given, mostly without proof. In particular, the correspondence between subalgebras and subgroups is discussed, along with homomorphisms, the exponential map, and the adjoint representation [25, 27, 33, 42, 55, 72].

2.1 Lie Groups

A *Lie group* G is a set which is both a group and a manifold and such that the group operations are C^∞, that is, the maps

$$G \times G \to G \qquad \text{given by} \qquad (g, h) \to gh$$

$$G \to G \qquad \text{given by} \qquad g \to g^{-1} \text{ are } C^\infty.$$

Examples. $Gl(d, R)$ is a Lie group under the standard operations of multiplying matrices. The group operations are given by rational functions in the coordinate variables, so that they are not only C^∞ but analytic. Since $Gl(d, R)$ may be considered as the group of nonsingular linear transformations on R^d, and every d-dimensional vector space V over R is isomorphic to R^d, the group of nonsingular linear transformations of V, $Gl(V)$, may be considered a Lie group isomorphic to $Gl(d, R)$. R^d is a Lie group under addition.

The circle $T^1 = S^1$ is a Lie group under the usual multiplication. T^e is then a Lie group by defining multiplication componentwise, and so also is $R^d \times T^e$. This is the most general Abelian Lie group. (See problem 13.)

In general, the product of Lie groups is a Lie group under componentwise multiplication, while a covering space of a connected

Lie group admits a Lie group structure such that the covering map is C^∞ and a homomorphism. In particular, every connected Lie group is locally isomorphic to a simply connected Lie group.

A *Lie subgroup* of a Lie group is a subgroup which is also a Lie group and a submanifold.

2.2 Lie Algebras

A *Lie algebra* is a vector space L for which is given a bilinear function from $L \times L$ to L, called *bracket*, and denoted by [,], which satisfies

(1) $[x, x] = 0$ for every $x \in L$

(2) the Jacobi identity: for every $x, y, z \in L$

$$[x, [y, z]] + [z, [x, y]] + [y, [z, x]] = 0.$$

A consequence of (1) is $[x, y] = -[y, x]$ and this relation implies (1) if L is not over a field of characteristic 2.

Examples

(1) If M is a C^∞ manifold, the global C^∞ vector fields form an infinite dimensional Lie algebra.

(2) The set of all linear transformations of R^d, $\mathfrak{gl}(d, R)$, is a Lie algebra under the bracket operation: $[A, B] = AB - BA$. We use similar notation, $\mathfrak{gl}(V)$, for the Lie algebra of linear transformations of a vector space V.

(3) For any vector space V a bracket may be defined by setting all brackets equal to 0. In this way we obtain the *Abelian* Lie algebra on V.

(4) If K and L are Lie algebras, then $K \oplus L$ is a Lie algebra with the bracket

$$[(x, y), (x', y')] = ([x, x'], [y, y']).$$

A *subalgebra* of a Lie algebra L is a subspace which is closed under the bracket operation of L. An *ideal* of a Lie algebra L is a subalgebra K such that for any $x \in L$ and $y \in K$, $[x, y] \in K$. A *homomorphism* of one Lie algebra into another is a linear transformation which preserves brackets. An *isomorphism* of Lie algebras is a homomorphism which is one-to-one onto.

The kernel of a homomorphism is an ideal. The image of a homomorphism is a subalgebra.

Let G be a Lie group. A *left invariant vector field* of G is a vector field which is fixed under the differentials of left translations, that is, if $L_g : G \to G$ is defined by $L_g(h) = gh$, then X is a left invariant vector field if $dL_g X(h) = X(gh)$ for every g, $h \in G$.

A left invariant vector field is globally defined and C^∞. The sum of two left invariant vector fields, the product of a left invariant vector field by a real number, and the bracket of two left invariant vector fields are again left invariant vector fields. A left invariant vector field is uniquely determined by its value at the identity of the group (cf. [25], pp. 102 and 103).

The *Lie algebra of G*, denoted by $\mathfrak{g}$, is the Lie algebra of left invariant vector fields.

According to the above remark, we may identify the vector space of the Lie algebra of G with the tangent space at the identity, namely, $X \in \mathfrak{g} \to X(e) \in G_e$.

The Lie algebra of $R^d \times T^e$ can be identified with the Abelian Lie algebra on R^{d+e}.

Locally isomorphic Lie groups have isomorphic Lie algebras. For example, R^2 and T^2 both have R^2 as Lie algebra, and similarly for any covering group of a Lie group. Also, the Lie algebra of the product of Lie groups is the direct sum of the Lie algebras.

General linear group. If we consider $Gl(d, R)$ as being the group of nonsingular linear transformations on R^d, then we get an isomorphism of the Lie algebra $\mathfrak{g}$ of $Gl(d, R)$ and $\mathfrak{gl}(d, R)$ as follows.

Let $\langle \, , \, \rangle$ be the usual inner product on R^d. For each v, $w \in R^d$ define a real-valued C^∞ function $f_{v,w}$ on $Gl(d, R)$ by

$$f_{v,w}(T) = \langle Tv, w \rangle, \qquad T \in Gl(d, R).$$

Then the map $J : \mathfrak{g} \to \mathfrak{gl}(d, R)$, given by $\langle J(X)\, v, \; w \rangle = X(e) f_{v,w}$ for every v, w, is the isomorphism.

Equivalently, we may do as follows: every element v of R^d may be considered as a C^∞ map $Gl(d, R) \to R^d$ if we set $v(T) = T(v)$. Then for each $X \in \mathfrak{g}$ we get a linear transformation $J(X)$ on R^d by defining $J(X)\, v = X(e)\, v$. Then J is a Lie algebra isomorphism $J : \mathfrak{g} \to \mathfrak{gl}(d, R)$. From now on we make the identification $J(X) = X$, that is, we shall write Xv for $X(e)\, v$.

On the other hand, if we consider $Gl(d, R)$ as matrices, and $\mathfrak{gl}(d, R)$ as the space of all $d \times d$ matrices, then $\mathfrak{g}$ may be identified with $\mathfrak{gl}(d, R)$ as follows: if $\{x_{ij}\}$ are the coordinate functions on $Gl(d, R)$, that is, $x_{ij}(g) = ij$th entry of g as a matrix, then for $X \in \mathfrak{g}$, $X_{ij} = X(e)(x_{ij})$.

Problem 1.   Show that: (a) $x_{ij} \circ L_g = \Sigma_p x_{ip}(g) \, x_{pj}$.

(b) $dL_g(D_{x_{hi}}(I)) \, x_{jk} = x_{jh}(g) \, \delta_{ik}$.

(c) If E_{ij} is the matrix with 1 in the ij position, 0 elsewhere, then the left invariant vector field X^{ij} corresponding to E_{ij} is

$$X^{ij} = \sum_k x_{ki} D_{x_{kj}}.$$

(d) Use this formula for X^{ij} to verify directly that the brackets are preserved under the identification of $\mathfrak{g}$ with $\mathfrak{gl}(d, R)$ [25].

2.3 Lie Group—Lie Algebra Correspondence

Theorem 1.   Let G be a Lie group. Then there is a one-one correspondence between the connected Lie subgroups of G and the subalgebras of the Lie algebra of G.

Outline of proof.   The correspondence is given as follows. If H is a Lie subgroup, then the left invariant vector fields on H can be extended uniquely to left invariant vector fields on G. The set of extensions form a subalgebra of the Lie algebra of G which is isomorphic to the Lie algebra of H.

The existence of the subgroup corresponding to a subalgebra is established by considering maximal integral manifolds of the involutive distribution obtained from the subalgebra. The one containing the identity of G is easily seen to be an abstract subgroup of G. It then follows trivially from problems 28 and 31 of Chapter 1 that it is a Lie subgroup (cf. [25], pp. 107–109).

For every Lie algebra the one-dimensional subspace generated by a nonzero element is a subalgebra. Thus we have:

Corollary.   For every $X \in \mathfrak{g}$, the Lie algebra of G, there is a one-dimensional subgroup with Lie algebra generated by X; that is, there is a curve $\gamma : R \to G$ such that

(1) $\gamma(s + t) = \gamma(s)\,\gamma(t)$

(2) $X(\gamma(s)) = \gamma_*(s)$.

$\gamma(R)$ is called the *one-parameter subgroup corresponding* to X. It is the integral curve of X which passes through e.

Orthogonal group. The group of all orthogonal transformations of R^d, $O(d)$, is a Lie subgroup of $Gl(d, R)$. Under the isomorphism J of 2.2 the Lie algebra of $O(d)$ is identified with the subalgebra $\mathfrak{o}(d)$ of $\mathfrak{gl}(d, R)$ consisting of all skew-symmetric transformations.

The fact that $O(d)$ is a Lie group follows from theorem 1 as applied to the component of the identity in $O(d)$, that is, the rotation group R_d or $SO(d)$, as a subgroup of the component of the identity of $Gl(d, R)$.

Lemma 1.   Let $J : G \to G$ by $J(g) = g^{-1}$. Then if $t \in G_g$,

$$dJ(t) = -dR_{g^{-1}} \circ dL_{g^{-1}}(t).$$

(R_g is right multiplication by g.)

Proof.   Let $D : G \to G \times G$ by $D(g) = (g, g)$,
$\qquad\qquad \beta : G \times G \to G$ by $\beta(g, h) = gh$.

Then $\beta \circ (1 \times J) \circ D = $ constant, so $d\beta \circ d(1 \times J) \circ dD = 0$. Let $t \in G_g$, then

$$
\begin{aligned}
0 &= d\beta \circ d(1 \times J) \circ dD(t) \\
&= d\beta \circ (d1(t) + dJ(t)) \\
&= d\beta(t + dJ(t)) \\
&= dR_{g^{-1}}(t) + dL_g(dJ(t)) \qquad \text{(theorem 1.2)} \qquad \text{QED}
\end{aligned}
$$

Corollary.   If $X \in \mathfrak{g}$, then dJX is right invariant.

There is a parallel formulation of the Lie algebra of a Lie group in terms of right invariant vector fields and dJ gives the connection with our formulation.

2.4 Homomorphisms

A *homomorphism* of one Lie group into another is a mapping which is both a homomorphism of the underlying groups and a C^∞ mapping of the underlying manifolds. We assume for the remainder of this section that our groups are connected.

If $j : G \to H$ is a homomorphism and t is a tangent at the identity of G, then it is easily verified that the left invariant vector fields corresponding to t and $dj(t)$, respectively, are j related. Thus j gives rise to a Lie algebra homomorphism $dj : \mathfrak{g} \to \mathfrak{h}$.

The converse of this last result is not true, unless we replace the concept of homomorphism by that of "local homomorphism" in which case the correspondence is one-to-one. [25, p. 112]. However, if G is simply connected, a local homomorphism can be extended to a homomorphism, so we obtain:

Theorem 2. Let G and H be Lie groups with G simply connected. Then the correspondence $j \leftrightarrow dj$ is one-to-one between homomorphisms of G into H and homomorphisms of $\mathfrak{g}$ into $\mathfrak{h}$.

The kernel of a homomorphism is a closed normal subgroup, and the kernel of the corresponding Lie algebra homomorphism is an ideal, and it is easily seen that this ideal belongs to the subgroup. Conversely, if H is a closed normal subgroup of G, then the set of left cosets G/H can be given a natural manifold structure in such a way that the projection $G \to G/H$ is a homomorphism of Lie groups. More generally we have:

Theorem 3. If H is a Lie subgroup of G, a necessary and sufficient condition for H to be normal is that its Lie algebra $\mathfrak{h}$ be an ideal in the Lie algebra $\mathfrak{g}$ of G. If, moreover, H is closed, the Lie algebra of G/H is naturally isomorphic to the factor algebra $\mathfrak{g}/\mathfrak{h}$ [25, pp. 115 and 124].

If G is simply connected and $\mathfrak{h}$ is an ideal of $\mathfrak{g}$, the Lie algebra of G, then by *Ado's theorem* there is a Lie group having Lie algebra $\mathfrak{g}/\mathfrak{h}$ [42]. Then by theorem 2 there is a homomorphism of G corresponding to the projection $\mathfrak{g} \to \mathfrak{g}/\mathfrak{h}$, from which it follows that the normal subgroup H belonging to $\mathfrak{h}$ is closed. Thus, if a Lie subgroup of a simply connected group is normal, then it is closed.

It is known that a closed subgroup of a Lie group is a Lie group (*Cartan's criterion*, [25, p. 135]).

2.5 Exponential Map

Let G be a Lie group, $X \in \mathfrak{g}$. Let γ_X be the integral curve of X starting at the identity. Then the *exponential map* $\mathfrak{g} \to G$ is the map which assigns $\gamma_X(1)$ to X; we write $\exp X = \gamma_X(1)$. Clearly the map $t \to \exp tX$ is just γ_X.

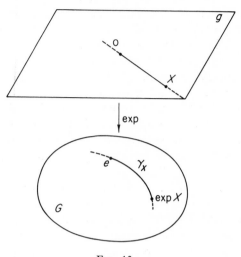

FIG. 13.

Commutativity with homomorphisms. If $j : G \to H$ is a homomorphism, then the diagram

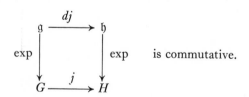

is commutative.

Proof. This follows immediately from the fact that $\gamma_{dj(X)} = j \circ \gamma_X$ for any $X \in \mathfrak{g}$.

From this and the following theorem we see that the exponential map gives the correspondence between subalgebras of $\mathfrak{g}$ and subgroups of G.

Theorem 4. The map $\exp$ is everywhere C^∞ and in a neighborhood of 0 in $\mathfrak{g}$, it is a diffeomorphism.

Proof.   Assuming for the moment that exp is C^∞ we first show that it is a diffeomorphism in a neighborhood of 0. It is sufficient, according to the inverse function theorem, to prove that d exp is onto at 0. If $s \in G_e$, then there is a left invariant vector field X such that $X(e) = s$. By definition, exp takes the ray generated by X (that is, the curve in $\mathfrak{g}$ given by $r \to rX$) into the integral curve of X through e. Hence d exp takes the tangent to this ray at 0 into the tangent to the integral curve of X at e, that is, into s.

We blame the differentiability of exp on theorems in differential equations which say that solutions depend in a C^∞ manner on parameters entering in a C^∞ manner into the functions defining a system of differential equations (see appendix). In this case we have the system of differential equations determined by a left invariant vector field X and X depends linearly on a system of linear coordinates of $\mathfrak{g}$. QED

Let $X_1, ..., X_d$ be a basis of $\mathfrak{g}$; then we identify $\mathfrak{g}$ with R^d as a manifold by the correspondence $\Sigma c_i X_i \leftrightarrow (c_1, ..., c_d)$. In a sufficiently small neighborhood U of 0 in $\mathfrak{g}$, exp is a diffeomorphism and hence can be taken as a coordinate map. The coordinates and coordinate neighborhood obtained in this way are designated as *canonical*.

Matrix exponential.   For matrix groups there is another exponential map, which turns out to be the same as the one already defined if we identify the Lie algebra of the matrix group with the right subalgebra of the Lie algebra of all matrices of the given order; namely, if A is a $d \times d$ matrix we define

$$e^A = \sum_0^\infty \frac{1}{k!} A^k \qquad (A^0 = I).$$

This series can be shown to converge in the norm induced on the space of $d \times d$ matrices through identification with R^{d^2}, and in fact, convergence is uniform on bounded sets.

Furthermore, it is evident that if $AB = BA$, then $e^{A+B} = e^A e^B$; in particular, $e^{(s+t)A} = e^{sA} e^{tA}$, s, t real numbers, and $e^A e^{-A} = e^0 = I$. Thus $e^A \in Gl(d, R)$ for every A, and the curve $s \to e^{sA}$ is a one-parameter subgroup; it is easy to see that the tangent at I of this one-parameter subgroup is naturally identified with $A = (d/du)(0)(e^{uA})$. In this way it is shown that the two exponential maps are essentially the same.

The following commutative diagram for $Gl(d, R)$ with the identification indicated by "$\approx$" will clarify relations.

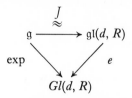

Problem 2. By showing that $B = \left(\begin{smallmatrix} -1/4 & 0 \\ 0 & -4 \end{smallmatrix}\right)$ has no square root, prove that exp does not map the Lie algebra of $Sl(2, R)(= 2 \times 2$ real matrices with determinant 1) onto $Sl(2, R)$.

Orthogonal group. A canonical coordinate neighborhood of the identity in $O(d)$ is obtained by taking the exponentials of skew-symmetric matrices lying in a sufficiently small neighborhood of 0. The coordinates for a skew-symmetric matrix may be taken as the elements of the matrix lying above the main diagonal. Thus the dimension of $O(d)$ is $\frac{1}{2}d(d-1)$.

For example, $O(2)$ is the one-dimensional group having as neighborhood of the identity the matrices

$$\exp \begin{pmatrix} 0 & \theta \\ -\theta & 0 \end{pmatrix} = \begin{pmatrix} \cos \theta & \sin \theta \\ -\sin \theta & \cos \theta \end{pmatrix},$$

and this correspondence is a diffeomorphism for $|\theta| < \pi$.

Problem 3. Let $A \in \mathfrak{gl}(d, R)$, $x \in R^d$, and define a C^∞ curve σ in R^d by $\sigma(s) = (\exp sA)\, x$. Prove that $\sigma_*(s) = A\,(\exp sA)\, x$, where tangents to R^d are identified with elements of R^d by the map $\Sigma\, a_i D_i(y) \to (a_1, ..., a_d)$.

Problem 4. Let $\langle\,,\,\rangle$ be the usual inner product on R^d and let σ_1, σ_2 be two C^∞ curves in R^d. Then a C^∞ function σ is defined by $\sigma(s) = \langle\sigma_1(s), \sigma_2(s)\rangle$. Using theorem 1.2 or otherwise, prove that

$$\sigma'(s) = \langle\sigma_{1_*}(s), \sigma_2(s)\rangle + \langle\sigma_1(s), \sigma_{2_*}(s)\rangle.$$

Problem 5. If $A \in \mathfrak{gl}(d, R)$ is skew-symmetric, $x \in R^d$, show that $\langle(\exp sA)\, x, (\exp sA)\, x\rangle$ is a constant function of s, thus equal to $\langle x, x\rangle$, the value at $s = 0$; thus $\exp sA$ is orthogonal for every s. Prove this directly using the map e above.

Problem 6. Define a complex Lie group. Examples are $Gl(d, C)$, C^d, and the complex torus $T^d(C) = C^d/D$, where D is a group generated by $2d$ real-linearly independent translations of C^d.

Problem 7. Show that a connected, compact, complex Lie group is Abelian (use a generalization of the maximum modulus theorem) [23].

Problem 8. Let $U(d) = \{g \in Gl(d, C) \mid g\bar{g}^t = I\}$. Show that $U(d)$ is a compact Lie group (the *unitary group*) but not a complex Lie group.

2.6 Representations

A *representation* of a Lie group G is a homomorphism of G into a matrix group. A *representation* of a Lie algebra $\mathfrak{g}$ is a Lie algebra homomorphism of $\mathfrak{g}$ into a Lie algebra of matrices. If instead of matrices we use the group, $Gl(V)$, [or Lie algebra $\mathfrak{gl}(V)$] of linear transformations of a vector space V, then we say we have a *representation on* V. In neither case do we exclude matrices with complex entries or vector spaces over the complex numbers. A *faithful representation* is a representation which is an isomorphism into; if a group has a faithful representation it is thus isomorphic to a subgroup of a matrix group.

Problem 9. (a) Show that $j : C \to Gl(2, R)$, $C =$ complex numbers, by $j(x + iy) = \left(\begin{smallmatrix} x & -y \\ y & x \end{smallmatrix}\right)$ is a faithful representation of the Lie group C^* of nonzero complex numbers with multiplication as operation. What is the subalgebra of $\mathfrak{gl}(2, R)$ corresponding to $j(C^*)$? What are the one-parameter groups of C^*?

(b) Show that the restriction of exp to the Lie algebra of $j(C^*)$ corresponds to the usual complex exponential function $e^{x+iy} = e^x (\cos y + i \sin y)$ and that C with addition as its operation is the simply connected covering group of C^* with $e^\cdot$ as the covering map.

(c) Construct a map $\phi : C^* \to S^1 \times S^1$ such that $\phi(e^{i\theta}) = (e^{i\theta}, 1)$ and ϕ is a Lie group homomorphism which covers $S^1 \times S^1$.

(d) Taking $1, i$ as the basis of C as a vector space over R, show that j corresponds to "multiplication by," that is, if $\psi(a + ib) = (a, b) \in R^2$ then the matrix $j(z)$ acts on $x \in R^2$ as $j(z) x = \psi(z\psi^{-1}(x))$.

Problem 10. Let $Q^* =$ nonzero quaternions. The *regular left re-*

presentation of Q^* on Q is the representation ϕ of Q^* on the real vector space Q given by $\phi(q)\,q' = qq'$ ($\phi(q) = $ "left multiplication by q").

(a) Show that with respect to basis 1, i, j, k (with the usual multiplication table) ϕ has matrix form, for $q = w + xi + yj + zk$,

$$\phi(q) = \begin{pmatrix} w & -x & -y & -z \\ x & w & -z & y \\ y & z & w & -x \\ z & -y & x & w \end{pmatrix}$$

(b) Show that $\phi(q)$ is orthogonal if and only if

$$|q| = w^2 + x^2 + y^2 + z^2 = 1.$$

(c) Compute $\det \phi(q)$ by showing $\phi(q)\,\phi(q)^t$ is a multiple of I and observing what the coefficient of w^4 is in $\det \phi(q)$.

Automorphisms. An *automorphism* of a Lie group G is an isomorphism of G onto itself. The set of all automorphisms of G form a group A. For every $j \in A$ we have the automorphism dj of the Lie algebra $\mathfrak{g}$ of G, and the diagram

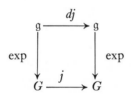

is commutative. Since dj is a nonsingular linear transformation of $\mathfrak{g}$ we have that the map $j \to dj$ takes A into the group of linear transformations of $\mathfrak{g}$, and it is evidently a homomorphism, since $d(j \circ k) = dj \circ dk$. Speaking loosely, we have a representation of A on $\mathfrak{g}$. In case G is connected this representation is faithful, a fact which follows from the remarks of 2.4. However, we prove this directly. Suppose j were in the kernel; this means that for every $X \in \mathfrak{g}$, $dj\,X = X$. However, since $j\,(\exp X) = \exp\,(dj\,X)$, j must then leave the image of exp fixed, and since this contains a neighborhood of the identity, which generates G, j is the identity automorphism.

The group A is actually a Lie group, but we do not show this here [*25*, pp. 137 and 138].

Adjoint representation. The set of inner automorphisms of G is a subgroup of A, that is, for every $x \in G$ we have the inner automorphism $j_x : y \to xyx^{-1}$. Moreover, the map $x \to j_x$ is a group homomorphism. The map $\mathrm{Ad} : G \to Gl(\mathfrak{g})$ defined by $\mathrm{Ad}(x) = dj_x$ is called the *adjoint representation of G*.

Proposition.   Ad is a representation of G on $\mathfrak{g}$.

Proof.   Ad is evidently a group homomorphism so that it suffices to show that it is C^∞, and in fact, to show that it is C^∞ in a canonical coordinate neighborhood.

First we note that for a fixed y in G the map $x \to j_x(y)$ is C^∞. In fact, it is the composition of maps involving group operations, which are obviously C^∞:

$$x \to (x, x) \to (x, xy^{-1}) \to (x, yx^{-1}) \to xyx^{-1}.$$

Now for y in a canonical coordinate neighborhood we have $y = \exp X$, and by commutativity, $j_x(y) = \exp(\mathrm{Ad}(x) X)$.

If we choose a basis $X_1, ..., X_d$ of $\mathfrak{g}$, then $\mathrm{Ad}(x)$ is given in terms of a matrix $(a_{ij}(x))$: $\mathrm{Ad}(x) X_j = \Sigma\, a_{ij}(x) X_i$. Now for $y = \exp(tX_j)$ we get $j_x(y) = \exp(\Sigma\, ta_{ij}(x) X_i)$, so that the canonical coordinates of $j_x(y)$ are $ta_{ij}(x)$, $i = 1, ..., d$, this being defined for t sufficiently small. Since $j_x(y)$ is C^∞ in x this means that $a_{ij}(x)$ is C^∞ in x for all i, j, that is, $\mathrm{Ad}(x)$ is C^∞.

The center.   The *center* of G is the set of all elements of G each of which commutes with every other element of G. It is clear that the kernel of the group homomorphism $x \to j_x$ is just the center of G, and since $j_x \to dj_x$ is faithful if G is connected, the center is the kernel of the adjoint representation. As such it is a closed subgroup, and being obviously normal, we have that the adjoint induces a faithful representation of the factor group $G/(\text{center of } G)$, whenever G is connected.

As a corollary to theorem 3.1 we shall see that the differential of the adjoint representation is the *adjoint representation* of $\mathfrak{g}$, which is given by $\mathrm{ad}(X)Y = [X, Y]$, that is, $d(\mathrm{Ad})(X)$ is the restriction of the Lie derivative with respect to X to $\mathfrak{g}$.

Problem 11.   Show that Q^* is the direct product of the positive real numbers and S^3, giving a *polar decomposition* of $Q^* = R^* \times S^3$. Since R^* is in the center of Q^* the adjoint representation has R^*

in its kernel and may be considered to have as representation space the tangent space to S^3 at the identity 1. Identify that space with the subspace of Q spanned by i, j, k and show that the adjoint representation on S^3 is then given by $\psi(q) P = qPq^{-1}$, $P = xi + yj + zk$. (For $q \in S^3$, $q^{-1} = \bar{q}$.) Identify the rest of the center of Q^* and the image of the adjoint representation on S^3 as a subgroup of $Gl(3, R)$.

Problem 12. Let X_1, X_2, X_3, X_4 be the left invariant vector fields on Q^* which are equal to D_i, $i = 1, 2, 3, 4$, at $1 = (1, 0, 0, 0)$. (Here we have identified Q with R^4.) Show that

$$X_1 = u_1 D_1 + u_2 D_2 + u_3 D_3 + u_4 D_4 ,$$

and find the corresponding formulas for X_2, X_3, X_4.

Problem 13. Show that an Abelian Lie group G has an Abelian Lie algebra (see problem 3.3) and hence by theorem 2, $G \approx R^d/D$, where D is a discrete subgroup. Therefore conclude that

$$G \approx R^e \times T^f, e + f = d \,[55, \text{pp. 83-86}]].$$

Problem 14. Show that a continuous homomorphism of a Lie group is C^∞, and hence show that a Lie group structure is a topological group invariant [25, p. 128].

Problem 15. Prove that an integral curve of a left invariant vector field is also an integral curve of a right invariant vector field.

Problem 16. Show that there is a neighborhood of the identity in a Lie group which contains no subgroups except the trivial one. Use this, problem 14, and the Peter-Weyl theorem [55, p. 99] to show that every compact Lie group has a faithful (C^∞) representation.

Remark on quotient spaces. In 2.4 we considered quotients of Lie groups by closed normal subgroups. Now let H be a closed subgroup of G. The H has the induced topology and the left cosets G/H have a natural manifold structure such that the projection $\pi\colon G \to G/H$ is C^∞, G acts as diffeomorphisms on G/H, and $f\colon G/H \to R$ is C^∞ if and only if $f \circ \pi$ is C^∞.

CHAPTER 3

Fibre Bundles

In the first section transformation groups are discussed and an interpretation of an important special case of the bracket operation is derived. The remainder of the chapter is devoted to principal and associated fibre bundles and reduction of the structural group, the treatment being from the point of view of transformation groups, although coordinate bundles are also defined [*4, 55, 66, 85*].

3.1 Transformation Groups

Let G be a Lie group and M a C^∞ manifold.

G *acts* (*differentiably*) *on* M *to the left* if there is a C^∞ map $\phi : G \times M \to M$, and we write $\phi(g, m) = gm$, satisfying the following conditions:

(a) For each $g \in G$, the map $g : M \to M$, given by $g(m) = gm$, is a diffeomorphism.

(b) For all $g, h \in G, m \in M, (gh)\, m = g(hm)$.

G is said to act *effectively* if $gm = m$, all $m \in M$, implies that $g = e = $ identity of G.

G acts on M *to the right* if (b) is replaced by:

(b)′ For all $g, h \in G, m \in M, (gh)\, m = h(gm)$. We also shall write $\phi: M \times G \to M$ in this case.

Every Lie group acts on itself to the left by left translation and by inner automorphism, while it acts on itself to the right by right translation.

If G acts on M, then to every $m \in M$ there corresponds a C^∞ map, denoted by m also, of G into M defined by: $m(g) = gm$.

38

G acts *transitively* to the left if for every m, $n \in M$, there is $g \in G$ such that $g(m) = n$. In this case, fixing some $m \in M$, let $H = \{g \in G \mid g(m) = m\}$, the *isotropy* group of m, then H is a closed subgroup of G and the map G/H (left cosets of H) $\to M$ defined by $gH \to gm$ is C^{∞}, one-to-one, onto. If G/H is compact, for example if G is compact, then this map is a homeomorphism.

Example. $Gl(d, R)$ acts differentiably to the left on R^d and on $R^d - \{0\}$. The action on $R^d - \{0\}$ is transitive; the isotropy group of $(1, 0, \ldots, 0)$ consists of matrices of the form $\left(\begin{smallmatrix} 1 & A \\ 0 & B \end{smallmatrix}\right)$, where

$$B \in Gl(d - 1, R), \ A \in R^{d-1},$$

0 is a column of $d - 1$ zeros. The subgroup H may be identified as the *semidirect product* of $Gl(d - 1, R)$ and R^{d-1}, that is, the multiplication is given by $(B, A)(B', A') = (BB', AB' + A')$. [In general, if group G acts as homomorphisms to the right on a group H, then the *semidirect product of G and H* is given by defining the products as $(g, h)(g', h') = (gg', (hg') h')$.]

Conversely, if H is a closed subgroup of G, then G acts transitively on G/H by $g(kH) = (gk)H$. A space with a transitive group of operators is called *homogeneous*.

If G acts on M, $\mathfrak{g}$ the Lie algebra of G, then we define a Lie algebra homomorphism λ of $\mathfrak{g}$ into a Lie algebra of vector fields on M, denoted by $\bar{\mathfrak{g}}$, as follows: if $X \in \mathfrak{g}$, then $(\lambda X)(m) = dm(X(e))$. We shall also write $\bar{X} = \lambda X$.

Problem 1. Prove that the one-parameter group of transformations of $\bar{X}$ is e^{tX}.

If G acts effectively, then λ is one-to-one.

G acts *freely* if the only element of G having a fixed point on M is the identity, that is, if for some $g \in G$ there is $m \in M$ such that $gm = m$, then $g = e$. If G acts freely, then the elements of $\bar{\mathfrak{g}}$ are nonvanishing vector fields on M. Moreover, if N is the *orbit of m*, that is, $N = \{gm \mid g \in G\}$, then for every $t \in N_m$ there is a unique $X \in \mathfrak{g}$ such that $\bar{X}(m) = t$, since $m : G \to M$ is a diffeomorphism ϕ of G with N and hence we may take the $X \in \mathfrak{g}$ such that $X(e) = d\phi^{-1}t$.

Problem 2. Prove that if $\bar{X}(m) = 0$ then $e^{tX}(m) = m$ for every t.

The differentials of the transformations making up a transformation group G, as they act on $\bar{\mathfrak{g}}$, are given as follows:

(a) If G acts to the left, then $dg(\bar{X}) = \overline{\text{Ad } g\, X}$.

(b) If G acts to the right, then $dg(\bar{X}) = \overline{\text{Ad } g^{-1}X}$.

Proof of (a). First we compute the composition of $g : M \to M$ and $g^{-1}m : G \to M : g \circ g^{-1}m(h) = g(hg^{-1}m) = ghg^{-1}m = m(j_g(h))$. Thus $(dg\ \bar{X})(m) = dg(\bar{X}(g^{-1}m)) = dg \circ d(g^{-1}m)\ X(e) = dm \circ dj_g(X(e)) = dm(Ad\ g\ X)(e)) = \overline{\text{Ad } g\ X}(m)$. QED

Let W be the linear space of C^∞ vector fields on M, and let $\mathscr{L}$ be a finite-dimensional subspace of W invariant under G, that is, for any $g \in G$, $dg(\mathscr{L}) \subset \mathscr{L}$. Hence, we have a representation[†] j of G in the group of nonsingular linear transformations of $\mathscr{L}$, that is, in $Gl(\mathscr{L})$; and hence, if $\mathfrak{gl}(\mathscr{L})$, the algebra of all linear transformations of $\mathscr{L}$, is regarded as the Lie algebra of $Gl(\mathscr{L})$, we have the commutative diagram

$$
\begin{array}{ccc}
\mathfrak{g} & \xrightarrow{\ dj\ } & \mathfrak{gl}(\mathscr{L}) \\
{\scriptstyle \exp}\big\downarrow & & \big\downarrow{\scriptstyle \exp} \\
G & \xrightarrow{\ j\ } & Gl(\mathscr{L})
\end{array}
$$

The following result has many uses.

Theorem 1. For every $X \in \mathfrak{g}$, $Y \in \mathscr{L}$, $dj(X)\,Y = -[\bar{X}, Y]$. In particular, $[\bar{X}, Y]$ is in $\mathscr{L}$.

Momentarily assuming this, we state and prove the corollary on the differential of the adjoint representation:

Corollary. If $X, Y \in \mathfrak{g}$, the Lie algebra of a Lie group G, then $\text{ad}(X)\,Y = [X, Y]$.

Proof. Let G act to the left on itself as follows: $g \in G$, $g : G \to G$ is defined by $g(h) = hg^{-1}$; that is, the transformation g is just $R_{g^{-1}}$, right multiplication by g^{-1}. Since $\text{Ad } g$ is $dL_g \circ dR_{g^{-1}}$, and X is

---

[†] Here we are assuming the action is a left action. In the case of a right action, obvious modifications are necessary in the definitions, although the statement of the theorem is the same.

invariant under left translations dL_g for every $X \in \mathfrak{g}$, we may apply the theorem with $\mathscr{L} = \mathfrak{g}$ and $j = \mathrm{Ad}$. We have to determine, for $X \in \mathfrak{g}$, the vector field $\bar{X}$ on G. For $m \in G = M$, we have that $m : G \to M$ as above is given by $m(g) = g(m) = mg^{-1}$, so $m = L_m \circ J$, where J is the inversing map. Thus $\bar{X}(m) = dL_m \circ dJ(X(e)) = dL_m(-X(e)) = -X(m)$, by lemma 2.1. Therefore $\bar{X} = -X$. Since $dj = d(\mathrm{Ad}) = \mathrm{ad}$, the theorem now gives the result.

Problem 3. Prove as a corollary to this: if G is Abelian, then $\mathfrak{g}$ is Abelian.

Proof of theorem. Assume the action is to the left. Let $X \in \mathfrak{g}$, so $\gamma : t \to \exp tX$ is a one-parameter group of diffeomorphisms of M, so there is a vector field Z on M which arises from differentiating functions along orbits. For a real-valued function f on M, and $m \in M$ we have

$$Zf(m) = \frac{d}{du}(0)(f \circ m \circ \gamma) = dm(\gamma_*(0))f,$$

so $Z(m) = dm(X(e)) = \bar{X}(m)$.

On the other hand, since the diagram above is commutative

$$\exp(t\, dj(X)) = \exp(dj(tX)) = j \exp(tX).$$

Thus, $dj(X)\, Y$ is the derivative at 0 of the curve $t \to (j \exp(tX))\, Y = d(\exp(tX))\, Y$ in $\mathscr{L}$. But for $m \in M$, $(d \exp tX(Y))(m) = d(\exp(tX))(Y(\exp(-tX)\, m))$, that is, the curve in M_m which we are differentiating is given by the values of Y along the curve $t \to \exp(t(-X))\, m$ pulled back to m by the action of the one-parameter group $t \to \exp(t(-X))$. Thus we are taking the Lie derivative with respect to $-\bar{X}$ (see 1.4), so by theorem 1.3 the conclusion follows.

3.2 Principal Bundles

A (C^∞) *principal fibre bundle* is a set (P, G, M), where P, M are C^∞ manifolds, G is a Lie group such that:

(1) G acts freely (and differentiably) to the right on P, $P \times G \to P$. For $g \in G$, we shall also write R_g for the map $g : P \to P$.

(2) M is the quotient space of P by equivalence under G, and the projection $\pi : P \to M$ is C^∞, so for $m \in M$, G is simply transitive on $\pi^{-1}(m)$.

(3) P is locally trivial, that is, for any $m \in M$, there is a neighborhood U of m and C^∞ map $F_U : \pi^{-1}(U) \to G$ such that F_U commutes with R_g for every $g \in G$ and the map of $\pi^{-1}(U) \to U \times G$ given by $p \to (\pi(p), F_U(p))$ is a diffeomorphism.

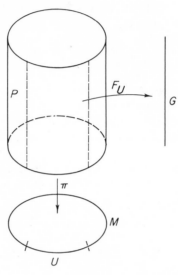

FIG. 14.

P is called the *bundle space*, M the *base space*, and G the *structural group*. For $m \in M$, $\pi^{-1}(m)$ is called the *fibre* over m. The fibres are diffeomorphic to G, in a special way via the map $p : G \to \pi^{-1}(\pi(p)) \subset P$, defined by $p(g) = R_g p$. We note that in terms of the F_U, the right action of G on P is given by right translation, that is, if $p \to (\pi(p,) F_U(p))$, then $pg \to (\pi(p), F_U(p) g)$. This follows from the fact that $F_U(pg) = F_U(p) g$.

We now give some examples of bundles.

Trivial (product) bundle. If G is a Lie group, M a manifold, then $P = M \times G$ provided with the right action of G on itself in the second factor, that is, $(m, g) h = (m, gh)$, is the bundle space of a principal bundle, the *trivial* bundle. A bundle is *isomorphic* to a trivial bundle if and only if there is a C^∞ cross section of π, that is, a C^∞ map $K : M \to P$ such that $\pi \circ K$ is the identity on M (cf. [85], pp. 25 and 36).

Bundle of bases. Let M be a C^∞ manifold and $B(M)$ the set of $(d+1)$-tuples $(m, e_1, ..., e_d)$, where $m \in M$ and $e_1, ..., e_d$ is a basis of M_m, and let $\pi : B(M) \to M$ be given by $\pi(m, e_1, ..., e_d) = m$. Then $Gl(d, R)$ acts to the right on $B(M)$ by: let $g \in Gl(d, R)$, viewed as a matrix, $g = (g_{ij})$; let $(m, e_1, ..., e_d) \in B(M)$, and define $R_g(m, e_1, ..., e_d) = (m, \Sigma g_{i1}e_i, ..., \Sigma g_{id}e_i)$. If $m \in M, (x_1, ..., x_d)$ a coordinate system defined in a neighborhood U of m, then we define F_U by: if $m' \in U$, $F_U(m', f_1, ..., f_d) = (dx_j(f_i)) = (g_{ij}) \in Gl(d, R)$. Thus the functions $y_i = x_i \circ \pi$ and $y_{ij} = x_{ij} \circ F_U$ give a coordinate system on $\pi^{-1}(U)$, where x_{ij} are the standard coordinates on $Gl(d, R)$ (see 2.2).

Using the C^∞ structure given to $B(M)$ by the local product representation (π, F_U), we see that $B(M)$ is the bundle space of a principal bundle, called the *bundle of bases* of M.

It is sometimes convenient to view $B(M)$ as the set of nonsingular linear transformations of R^d into the tangent spaces of M, that is, we identify $p = (m, e_1, ..., e_d)$ with the map $p : (r_1, ..., r_d) \to \Sigma r_i e_i$. When this is done it is natural to consider $Gl(d, R)$ as the nonsingular linear transformations of R^d, for we have: $pg(r_1, ..., r_d) =$

$$\sum_{i,j} r_i g_{ji} e_j = \sum_j \left(\sum_i r_i g_{ji} \right) e_j = p(g(r_1, ..., r_d)),$$

that is, pg (as a map) $= p$ (as a map) $\circ g$.

Problem 4. If $b = (m, e_1, ..., e_d) \in B(M)$ is in the coordinate neighborhood $\pi^{-1}(U)$ show that $d\pi(D_{y_i}(b)) = \Sigma_j y_{ji}^{-1}(b)e_j$, where $(y_{ij}^{-1}(b))$ is the inverse of the matrix $(y_{ij}(b))$.

Homogeneous spaces. If G is a Lie group, H a closed subgroup, then there is a principal bundle with base space G/H (left cosets), bundle space G, and structure group H such that $\pi : G \to G/H$ is the canonical map and the right action is given by $(g, h) \to gh$ (see [85], p. 33).

Examples (1) Let $R - \{0\} = R^*$ act on $R^{d+1} - \{0\}$ by scalar multiplication. Then this action is differentiable, free, and simply transitive on orbits. The orbit space is P^d, d-dimensional projective space, so

$$(R^{d+1} - \{0\}, R^*, P^d) \qquad \text{is a principal bundle.}$$

(2) If we use the positive reals R^+ instead of R^* we get

$$(R^{d+1} - \{0\}, R^+, S^d) \qquad \text{is a principal bundle.}$$

(3) If we use C^* and $C^{d+1} - \{0\}$, C = complex numbers, then we get complex projective space CP^d as base space:

$$(C^{d+1} - \{0\}, C^*, CP^d) \qquad \text{is a principal bundle.}$$

Problem 5.   Let (P, G, M) be a principal fibre bundle with P connected. Then if G_0 is the component of the identity in G, there is a unique principal fibre bundle $(P, G_0, \tilde{M})$ such that the action of G_0 on P is the same and $\tilde{M}$ is a connected covering space of M (cf. examples 1 and 2 above).

We give an alternate approach to principal fibre bundles.

Principal coordinate bundles.   Let (P, G, M) be a principal bundle, and let $\{U_i\}$ be an open covering of M such that $\pi^{-1}(U_i)$ can be represented as a product space via the function $F_i : \pi^{-1}(U_i) \to G$. For i, j such that $U_i \cap U_j \neq 0$, we define a map $G_{ji} : U_i \cap U_j \to G$ as follows: if $m \in U_i \cap U_j$, let $p \in \pi^{-1}(m)$, and put $G_{ji}(m) = F_j(p)(F_i(p))^{-1}$. G_{ji} measures how much the cross section over U_i corresponding to $U_i \times \{e\}$ under the product structure given by F_i differs from the cross section over U_j determined similarly by F_j. We want to show that the definition of $G_{ji}(m)$ is independent of the choice of p. If $p' \in \pi^{-1}(m)$ also, then since $\pi^{-1}(m)$ is the orbit of p, there is $g \in G$ such that $p' = pg$, so we have

$$F_j(p') F_i(p')^{-1} = F_j(pg) F_i(pg)^{-1} = F_j(p) g (F_i(p) g)^{-1}$$
$$= F_j(p) F_i(p)^{-1},$$

as desired. The functions satisfy the further properties:

$$G_{ki}(m) = G_{kj}(m) \, G_{ji}(m) \qquad \text{for} \qquad m \in U_i \cap U_j \cap U_k. \tag{*}$$

The functions G_{ji} are called the *transition functions* corresponding to the covering $\{U_i\}$, and in fact, with this covering they define a principal coordinate bundle in the sense of Steenrod. Hence, our definition of bundle is an equivalence class of coordinate bundles, which is another way of saying that principal coordinate bundles are equivalent if and only if their right actions agree. Therefore, we have, by [85, p. 14], that any set of functions G_{ji} defined for a covering $\{U_i\}$, satisfying (*), uniquely determine a principal bundle whose transition functions relative to the covering $\{U_i\}$ are the G_{ji}.

Problem 6. If $\phi: N \to M$ is a C^∞ map, (P, G, M) a principal fibre bundle, then let $\tilde{P} = \{(n, p) \in N \times P \mid \phi(n) = \pi(p)\}$.

(a) Show that $\tilde{P}$ is a submanifold of $N \times P$ under the inclusion map.

(b) Show that $(\tilde{P}, G, N)$ becomes a principal fibre bundle if we define right action by $(n, p)\, g = (n, pg)$.

$(\tilde{P}, G, N)$ is called the *bundle induced by* ϕ *and* (P, G, M).

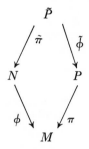

(c) Show that transition functions for (P, G, N) may be taken as $G_{ij} \circ \phi$, so that we could define the coordinate bundle directly.

The vector fields of $\bar{\mathfrak{g}}$. If (P, G, M) is a principal bundle, then since G acts freely and effectively we have an isomorphism $\lambda: \mathfrak{g} \to \bar{\mathfrak{g}} =$ Lie algebra of vector fields on P, which consists of nonvanishing vector fields. By remark (*b*) in 3.1 we have $dR_g(\lambda(X)) = \lambda(\mathrm{Ad}(g^{-1})\, X)$, where $g \in G$, $X \in \mathfrak{g}$.

Problem 7. For the real coordinates $u_1, ..., u_{2d+2}$ on $C^{d+1} - \{0\}$ ($\{z_j = u_{2j-1} + iu_{2j}\}$ is the dual basis to the standard complex basis of C^{d+1}) find the expression for $\lambda(aX_1 + bX_2)$, where X_1 and X_2 are the left invariant vector fields on C^*: $X_1 = u_1 D_1 + u_2 D_2$, $X_2 = -u_2 D_1 + u_1 D_2$ and λ is the isomorphism associated with $(C^{d+1} - \{0\}, C^*, CP^d)$.

3.3 Associated Bundles

Let (P, G, M) be a principal fibre bundle, and let F be a manifold on which G acts to the left. We define the *fibre bundle associated to*

(P, G, M) *with fibre* F (it also depends on the action) as follows. Let $B' = P \times F$, and consider the right action of G on B' defined by $(p, f) g = (pg, g^{-1}f)$, where $p \in P, f \in F, g \in G$. Let $B = B'/G$, the quotient space under equivalence by G, then B is the bundle space of the associated fibre bundle. We have the following structure. The projection $\pi': B \to M$ is defined by: $\pi'((p, f) G) = \pi(p)$. If $m \in M$, take U a neighborhood of m as in 3.2(3), with $F_U : \pi^{-1}(U) \to G$. Then we have $F_U' : \pi'^{-1}(U) \to F$ given by

$$F_U'((p,f) G) = F_U(p) f,$$

so that $\pi'^{-1}(U)$ is homeomorphic to the product $U \times F$, and hence we define B as a manifold by requiring these homeomorphisms to be diffeomorphisms. Note that now $\pi' \in C^\infty$, and also the natural projection $B' \to B$ is C^∞.

Associated coordinate bundle. If we define transition functions G_{ji}' for the associated bundle (B, F, G, M) analogously to the definition in the principal bundle case, we have, for a covering $\{U_i\}$ admitting functions F_i', for $m \in U_i \cap U_j$, $(p, f) G \in \pi'^{-1}(m)$:

$$G_{ji}'(m) = F_j(p)(F_i(p))^{-1} = G_{ji}(m).$$

We therefore have that (B, F, G, M) is the equivalence class of the coordinate bundle associated to the principal coordinate bundle defined by the transition functions G_{ji}, in the sense of Steenrod [85, Part I, §9].

Examples

(1) Let (P, G, M) be as above, and let G act on itself by left translation. Then (P, G, M) is the bundle associated to itself with fibre G.

(2) *Tangent bundle.* Consider the bundle of bases $B(M)$. (Note that we shall often denote a bundle by its bundle space alone.) Now by definition $Gl(d, R)$ is the group of nonsingular linear transformations of R^d, and hence acts on R^d to the left. The bundle space of the associated bundle with fibre R^d is denoted by $T(M)$, and the bundle is called the *tangent bundle* to M. $T(M)$ can be identified with the space of all pairs (m, t), where $m \in M, t \in M_m$ [the "m" in the pair

is actually redundant, as it is in the $(d + 1)$-tuples making up $B(M)$, but is inserted for convenience], as follows:

$$((m, e_1, ..., e_d), (r_1, ..., r_d)) \, Gl(d, R) \to (m, \Sigma \, r_i e_i),$$

or, if we regard $B(M)$ as the set of maps $p : R^d \to M_m$, then for $x \in R^d$ this identification is $(p, \, x) \, Gl(d, R) \to (m, px)$, where $m = \pi(p)$. With this later formulation it is easy to see that the identification is well defined, for if $(p', \, x') \, Gl(d, R) = (p, \, x) \, Gl(d, R)$, then there is $g \in Gl(d, R)$ such that $p' = pg$, $x' = g^{-1}x$, and hence, $p'x' = (pg)(g^{-1}x) = px$, since as a map, pg is the composition of p and g.

Hence we may view the fibre of $T(M)$ above $m \in M$ as the linear space of tangents at m, that is, as M_m, and $T(M)$ itself as the union of all the tangent spaces supplied with a manifold structure. Furthermore, under this identification, the coordinates of $T(M)$ may be easily exhibited; namely, let U be a coordinate neighborhood in M, with coordinates $x_1, ..., x_d$. We define coordinates $y_1, ..., y_{2d}$ on $\pi'^{-1}(U)$ as follows: if $(m, t) \in \pi'^{-1}(U)$,

$$\left. \begin{aligned} y_i(m, t) &= x_i(m) \\ y_{d+i}(m, t) &= dx_i(t) \end{aligned} \right\} \quad i = 1, ..., d.$$

A C^∞ vector field may then be regarded as a cross section of π'. In particular, the trivial vector field (values all 0) gives an imbedding of M as a submanifold of $T(M)$.

Problem 8. Prove that if γ is a C^∞ curve in M, then γ_* is a C^∞ curve in $T(M)$.

(3) *Tensor bundles.* When we replace R^d in example (2) by a vector space constructed from R^d via multilinear algebra, that is, the tensor product of R^d and its dual with various multiplicities or an invariant subspace thereof, we get a *tensor bundle*. A cross section which is C^∞ on an open set is called a C^∞ *tensor field*, and is given type numbers according to the number of times R^d and its dual occur. The group of a tensor bundle is $Gl(d, R)$; it acts on the factors of the tensor product independently, and on R^d as with the tangent bundle, on the dual via the transpose of the inverse: if $v \in$ the dual, $x \in R^d$, $g \in Gl(d, R)$, then $gv(x) = v(g^{-1}x)$.

Frequently the bundle $B(M)$ has its structural group reduced to a subgroup (see 3.4) in which case more subspaces of tensor products of R^d and its dual may become invariant, leading to tensor bundles of different sorts. For example, this is the case when M has an almost complex structure (cf. problem 11).

(4) *Vector bundles.* These are bundles in which the fibre is a vector space and the structural group is a subgroup of the general linear group of that vector space. The tensor bundles are a special case. They are frequently defined with no explicit mention made of the structural group by giving the bundle space as the union of vector spaces, all of the same dimension, each associated to an element of the base space, and defining the manifold structure via smooth, linearly independent, spanning cross sections over a covering system of coordinate neighborhoods. For example, we essentially did this for $T(M)$ when we exhibited the coordinates: in that case the cross sections were the coordinate vector fields D_{x_i}.

Another example of this type is the *quotient space bundle* of an imbedding, which usually is considered in the case of Riemannian manifolds as the *normal bundle* (the Riemannian metric, defined in Chapter 7, is employed to get a uniform choice of representatives for the quotient spaces). This vector bundle may be defined as follows: Let $i : N \to M$ be the imbedding of the submanifold N in M. The fibre over $n \in N$ is the quotient space $M_{i(n)}/di(N_n)$, and the bundle space is the union of these fibres, so we may consider the bundle space as the collection of pairs $(n, t + di(N_n))$, where $t \in M_{i(n)}$. To get coordinates we first take a coordinate system in M at $i(n)$, say $x_1, ..., x_d$, and we may assume that $x_j \circ i = y_j$, $j = 1, ..., d'$ give a coordinate system at n, and that $X_j(n) = D_{x_j}(i(n)) + di(N_n)$ are linearly independent for $j = d' + 1, ..., d$. Then for some neighborhood U of n, $X_j(n')$, $j = d' + 1, ..., d$ are still linearly independent, so there is a dual basis $V_j(n')$. Then for $(n', X) \in \pi'^{-1}(U)$, where π' is the projection from the bundle space to N, we define coordinates $z_1, ..., z_d$ by:

$$z_j(n', X) = y_j(n') \qquad \text{if} \quad j = 1, ..., d'$$
$$z_j(n', X) = V_j(n')(X) \qquad \text{if} \quad j = d' + 1, ..., d.$$

The group of this bundle may be taken to be $Gl(d - d', R)$.

(5) *Grassmann bundles.* The set of e-dimensional subspaces of R^d can be given a manifold structure so that $Gl(d, R)$ acts in the obvious manner as a differentiable transformation group on the left. The bundles associated to $B(M)$ by this action are called the (unoriented)

Grassmann bundles of M; the bundle space may be regarded as the collection of e-planes in the tangent spaces of M. A C^∞ e-dimensional distribution is a C^∞ cross section of this bundle.

The action of a principal bundle on an associated fibre. Let (P, G, M) be a principal bundle and B an associated bundle with fibre F. Then the quotient projection $P \times F \to B$ defines, by restriction of the first variable, for every $p \in P$ a C^∞ map $p : F \to B$, namely, $p(f) = (p, f) G$. It satisfies $p(gf) = (pg)f$ for every $g \in G$. We have already seen this in the case of $B(M)$ and $T(M)$ (cf. [*85*], Part I, §8.9).

Remark. An associated bundle is *trivial* if its principal bundle is trivial. This is not equivalent to the existence of a cross section of the associated bundle, but implies the existence of a family of cross sections with pairwise disjoint ranges which fill the associated bundle.

Problem 9. Verify the above, and also show that for a vector bundle with fibre R^e triviality is equivalent to the existence of e cross sections linearly independent at each point.

It is well known [*85*] that the tangent bundle to a differentiable manifold admits a nonzero cross section if and only if the Euler characteristic is zero; for example, if the manifold is compact and odd-dimensional. Hence all odd-dimensional spheres have such cross sections, although it is a deep result of Milnor and Kervaire [*14*] that only S^1, S^3, S^7 have trivial tangent bundles (i. e., are *parallelizable*).

3.4 Reduction of the Structural Group

Let (P, G, M) be a principal bundle. We assume G is separable. Let H be a subgroup of G, then in the sense of Steenrod, the structural group G is *reducible to the subgroup* H if and only if there exists a coordinate bundle in the equivalence class determined by (P, G, M) whose transition functions take their values in H, that is, if and only if there exists a covering $\{U_i\}$ whose G_{ji} satisfy $G_{ji}(U_i \cap U_j) \subset H$.

In terms of the right action $P \times G \to P$, this definition can be formulated in the following way ([*66*], p. 20).

Let $(P, G, M), (P', G', M')$ be principal bundles. A *bundle map* $f : (P, G, M) \to (P', G', M')$ is a set of C^∞ maps (f_P, f_G, f_M) between the obvious pairs, such that f_G is a homomorphism, and the following relations are satisfied:

$$f_M \circ \pi = \pi' \circ f_P$$
$$f_P \circ R_g = R_{f_G(g)} \circ f_P \qquad \text{for every} \qquad g \in G.$$

Now we may state the second definition as a theorem.

Theorem 2. If (P, G, M) is a principal bundle, H a subgroup of G, then G is reducible to H if and only if there exists a principal bundle (P', H, M) which admits a bundle map $f: (P', H, M) \to (P, G, M)$ such that f_M is the identity on M, f_P is one-to-one, and f_G is the inclusion map $H \subset G$ (Proof omitted.)

Steenrod proves [85, p. 59] that if (P, G, M) is a principal bundle, H a maximal compact subgroup of G, then G can be reduced to a bundle with structural group H. In particular, every principal bundle with $Gl(d, R)$ as structural group [for example, $B(M)$] can be reduced to a bundle with structural group the orthogonal group $O(d)$. We shall return to this when we consider Riemannian metrics on a manifold.

The reduction of a principal bundle induces the reduction of associated bundles, in an obvious sense, since we have given the definition in terms of transition functions only.

Problem 10. *Complex manifolds.* Let M be a complex manifold, $\mathscr{F}$ = complex valued differentiable functions defined on a neighborhood of $m \in M$, $\mathscr{F}_a$ = holomorphic functions in $\mathscr{F}$, $\bar{\mathscr{F}}_a$ = conjugates of functions in $\mathscr{F}_a$ (conjugate holomorphic functions), $\mathscr{F}_r$ = real valued functions in $\mathscr{F}$, $\mathscr{T}_m$ = complex linear derivations of $\mathscr{F}$, $\mathscr{M}_m$ complex linear extensions of M_m, $\mathscr{H}_m$ = the annihilator of $\bar{\mathscr{F}}_a$ in $\mathscr{T}_m$. For $t \in \mathscr{T}_m$, $f \in \mathscr{F}$, define $\bar{t}f = \overline{t\bar{f}}$. Show that

(a) $\mathscr{T}_m = \mathscr{M}_m + i\mathscr{M}_m$ (direct sum).

(b) $\bar{t} \in \mathscr{T}_m$ for every $t \in \mathscr{T}_m$.

(c) $\mathscr{M}_m$ = all $t \in \mathscr{T}_m$ such that $t\mathscr{F}_r \subset R$.

(d) $\bar{\mathscr{H}}_m$ = the annihilator of $\mathscr{F}_a$.

(e) $t = \bar{t}$ if and only if $t \in \mathscr{M}_m$.

(f) $\mathscr{H}_m \cap \bar{\mathscr{H}}_m = 0$.

(g) $\mathscr{T}_m = \mathscr{H}_m + \bar{\mathscr{H}}_m$.

(h) If $t \in \mathcal{M}_m$, then the decomposition of t from (g) is of the form $t = h + \bar{h}$, where $h \in \mathcal{H}_m$, i.e., $\mathcal{M}_m = \{h + \bar{h} \mid h \in \mathcal{H}_m\}$. This defines a one-to-one real linear map $P : \mathcal{M}_m \to \mathcal{H}_m$, $t \to h$.

Let $j : \mathcal{T}_m \to \mathcal{T}_m$ be multiplication by i. Define $J = P^{-1}jP$.

(i) $J^2 = -$identity.

(j) Compute J in terms of real coordinate vector fields which come from the real and imaginary parts of a complex coordinate system.

(k) J is defined on $T(M)$ and is a bundle map.

An *almost complex structure* on a manifold M is a bundle map $J : T(M) \to T(M)$ such that

(1) $J(M_m) = M_m$ for every $m \in M$.

(2) $J^2 = -$identity on each M_m.

The J obtained in the above problem is called the *complex structure* of the complex manifold M. An almost complex structure will be called *a complex structure* only if it is obtained in this way.

Problem 11. (a) If M has an almost complex structure, then M has even dimension.

(b) M has an almost complex structure if and only if the group of the bundle of bases can be reduced to $Gl(d/2, C)$ represented in $Gl(d, R)$ as matrices of the form $\begin{pmatrix} A & -B \\ B & A \end{pmatrix}$ (which corresponds to $(A + iB) \in Gl(d/2, C)$).

Every 2-dimensional orientable differentiable manifold admits a complex structure, so, of course, every 2-dimensional manifold which admits an almost complex structure admits a complex structure. The latter result is not true for higher dimensional manifolds [21].

If M has an almost complex structure then $M_m + iM_m$ has a direct sum decomposition $\mathcal{H}_m + \overline{\mathcal{H}}_m$ such that $\mathcal{H}_m = $ all $x + iJx$ for $x \in M_m$. We say that C^∞ vector field X belongs to $\mathcal{H}$ if $X(m) \in \mathcal{H}_m$ for every m, and then write $X \in \mathcal{H}$. It is not in general true that if $X \in \mathcal{H}$, $Y \in \mathcal{H}$ then $[X, Y] \in \mathcal{H}$. However, if J is a complex structure the $\mathcal{H}_m$ agrees with our previous definition and $[\mathcal{H}, \mathcal{H}] \subset \mathcal{H}$; in fact, this latter condition has been shown to be also a sufficient condition that J be a complex structure [64].

Problem 12. The maximal compact subgroups of R^* and C^* are $\{1, -1\} = S^0$ and $\{e^{i\theta}\} = S^1$, respectively. Show that the reduction of the principal bundles of examples (1) and (3) in 3.2 to these sub-groups gives principal bundles: $(S^d,\ S^0,\ P^d)$ and $(S^{2d+1},\ S^1,\ CP^d)$.

Carry out the same construction to get principal bundles over quaternionic projective spaces QP^d: $(S^{4d+3},\ S^3, QP^d)$.

In the case $d = 1$ the bundle projections become the Hopf maps $S^3 \to S^2 = CP^1,\ S^7 \to S^4 = QP^1$.

CHAPTER 4

Differential Forms

Differential forms are defined via Grassmann algebras, and the intrinsic formula for the exterior derivative is derived. Frobenius' theorem, vector-valued forms, and forms on complex manifolds are also discussed [*24, 25, 29, 33, 36, 66, 83*]. For other topics, particularly the use of differential forms in the study of topological invariants, the reader is referred to [*12, 30, 78*].

4.1 Introduction

In the last chapter the concept of a tensor was briefly mentioned, and differential forms, the subject of the present chapter are just special types of tensors. However, we shall initially introduce differential forms here by means of the more explicit formulation in terms of Grassmann algebras and shall then return briefly to the tensorial approach (4.5).

If f is a C^∞ function on M, then we notice that to every $m \in M$ there corresponds the differential of f at m, 1.3, which is a linear functional on M_m, and this correspondence is smooth in the following sense. Let X be a C^∞ vector field on M, then $df(X)(m) = df_m X(m) = Xf(m)$ defines a C^∞ function Xf on M. Such a smooth assignment of linear functions is called a differential 1-form, although not every differential 1-form arises as the differential of a C^∞ function. However, before pursuing this subject further, we must develop some machinery, namely, Grassmann algebras.

4.2 Classical Notion of Differential Form

A differential form at m is something which in terms of a coordinate system can be expressed as $\sum a_{i_1 \ldots i_p} \, dx_{i_1} \cdots dx_{i_p}$, where the summation is over all ordered subsets $(i_1, ..., i_p)$ of $\{1, ..., d\}$, and the $a_{i_1...i_p}$ are real numbers.

So a strict definition will involve some kind of multiplication of differentials and then linear combination of these products. Thus $M_m{}^*$ will be imbedded in an algebraic system which has both multiplication and vector operations. Furthermore, this shall be done so that if $(y_1, ..., y_d)$ is a second coordinate system at m then the expression $\Sigma \, b_{i_1...i_p} \, dy_{i_1} \cdots dy_{i_p}$, obtained from the other by the usual rules for change of variable in multiple integrals, will be just the expansion of the same algebraic object in terms of the new basis.

The required algebraic system for this is a Grassmann algebra, which we now discuss.

4.3 Grassmann Algebras

Let F be a field and V a finite-dimensional vector space over F of dimension d. A *Grassmann algebra over* V is a set G such that

(i) G is an associative algebra with identity e over F.

(ii) G contains V.

(iii) Every element $v \in V$ satisfies $v^2 = 0$.

(iv) G has dimension 2^d.

(v) G is generated by e and V, that is, every element of G is a sum of products of scalar multiples of e and elements of V.

Notice that e is not in V because $e^2 = e \neq 0$, while by (iii) $v^2 = 0$ for all $v \in V$. Also, if $u, v \in V$, then $uv = -vu$. This is shown by the standard polarization trick:

$$0 = (u + v)^2 = u^2 + uv + vu + v^2 = uv + vu.$$

Property (iv) is a short but poor way of stating that there are no more relations among the elements of G than those which follow from (iii).

To each basis $e_1, ..., e_d$ of V there corresponds a basis of G. The elements of this basis of G are in one-to-one correspondence with the subsets of $\{1, ..., d\}$ as follows:

(a) If the subset is $\phi = $ empty set, we let $e_\phi = e$.

(b) If the subset is $s = \{i_1, ..., i_p\}$ with $i_1 < \cdots < i_p$, let $e_s = e_{i_1} \cdots e_{i_p}$.

Note that there are exactly 2^d subsets, so that in order to show that the e_s's are a basis it is enough to prove the:

Lemma 1. The e_s's span G.

Proof. By (v) above we have for any $g \in G$

$$g = a_0 e_\phi + \sum \text{(products of elements of } V)$$

$$= a_0 e_\phi + \sum \text{(coefficients in } F) \text{ (products of } e_i\text{'s)}.$$

But by using the fact that elements of V anticommute we can write any product of e_i's in increasing order and then by (iii) all except the e_s's are 0.

An element $g \in G$ is *homogeneous of degree* p if it can be written as the sum of products of exactly p elements of V, or, equivalently, $g = \Sigma_{s \in P} a_s e_s$, where $a_s \in F$, P = subsets of $\{1, ..., d\}$ which have exactly p elements.

The set of all such $g \in G$ form a linear subspace G^p of G which is of dimension $\binom{d}{p}$. Clearly, $G^p G^q \subset G^{p+q}$. Moreover, by anticommutativity it is easily seen that if $g \in G^p$, $h \in G^q$ then $gh = (-1)^{pq} hg$.

An element g of G which is homogeneous of degree p is *decomposible* if there exist $v_1, ..., v_p \in V$ such that $g = v_1 ... v_p$. Otherwise g is called *indecomposible*.

Problem 1. If dim $V \leqslant 3$, then every homogeneous element is decomposible. If dim $V > 3$, then if v_1, v_2, v_3, v_4 are linearly independent, $v_1 v_2 + v_3 v_4$ is indecomposible.

In general, the decomposible elements form an (nonlinear) algebraic variety in G^p.

Problem 2. Prove that for $g \in G^2$ there exist linearly independent $v_1, ..., v_{2k} \in V$ such that $g = v_1 v_2 + v_3 v_4 + \cdots + v_{2k-1} v_{2k}$; that if the characteristic of F is not a prime $\leqslant \frac{1}{2} d$, then k is the largest integer such that $g^k \neq 0$, and in this case g^k is decomposible.

We say that g has *rank* $2k$.

Problem 3. If dim $V = 4$ and the characteristic of F is not 2 then g is decomposible if and only if $g^2 = 0$ and g is homogeneous. In general, if dim $V = 4$ and $g \in G^{d-1}$, then g is decomposible.

Problem 4. Let $x \neq 0$, $x \in V$, $g \in G$. Prove $xg = 0$ if and only if there is $h \in G$ such that $g = xh$.

Remarks. (1) If $v_1, ..., v_p \in V$, then $v_1 ... v_p \neq 0$ if and only if $v_1, ..., v_p$ are linearly independent.

Proof. If $v_1, ..., v_p$ are linearly independent, then we may take them as part of a basis for V; so that $v_1 ... v_p = e_s \neq 0$, $s = \{1, ..., p\}$.

If $v_1, ..., v_p$ are linearly dependent, say, $v_1 = \Sigma_{i=2}^{p} a_i v_i$, then by the distributive law and anticommutativity we have

$$v_1 \cdots v_p = \sum_{i=2}^{p} \pm a_i v_2 \cdots v_i^2 \cdots v_p = 0.$$

(2) Any two Grassmann algebras over V are essentially the same, that is, if G and H are two such then there is an isomorphism of G onto H which leaves each element of V fixed. In fact we need only choose a basis of V and map the e_s's for G into the corresponding ones for H.

(3) If T is a linear transformation of a vector space V into a vector space W, and $G(V)$, $G(W)$ are the Grassmann algebras over V, W, respectively, then there is a unique extension of T to a homomorphism of $G(V)$ into $G(W)$, denoted by T_h, and T_h satisfies $T_h(G(V)^p) \subset G(W)^p$.

Proof. Uniqueness follows immediately from the facts that $T(e)$ must equal e, and that $G(V)$ is generated by e and V. By linearity, T_h can be extended to all of $G(V)$ after defining

$$T_h(v_1 \cdots v_p) = T(v_1) \cdots T(v_p) \qquad \text{for} \qquad v_1, ..., v_p \in V,$$

and such a T_h is clearly a homomorphism.

Problem 5. If $T : V \to W$, $S : W \to X$ are linear transformations of vector spaces, show that $(ST)_h = S_h T_h$.

Problem 6. If $T : V \to V$ is a linear transformation, then it admits a unique extension to a derivation of $G(V)$ into $G(V)$, that is, a map T_d such that if $g, h \in G(V)$ then $T_d(gh) = T_d(g)\, h + g T_d(h)$.

Problem 7. If $T, S : V \to V$, show that $[S_d, T_d]$ is a derivation and that $[S, T]_d = [S_d, T_d]$, that is, $(ST)_d - (TS)_d = S_d T_d - T_d S_d$. Give an example to show that $S_d T_d$ is not always a derivation.

If V is a d-dimensional vector space and G its Grassmann algebra, then G^d is one dimensional. A linear transformation of G^d into itself is simply scalar multiplication by an element of F. If S is the transformation then we denote this scalar by $k(S)$.

Now let T be a linear transformation of V into itself. By (3) and problem 6 there are extensions T_h and T_d of T to G which are a homomorphism and a derivation, respectively. In particular, T_h and T_d are linear transformations of G^d. We define

$$\text{determinant of } T = \det T = k(T_h),$$

$$\text{trace of } T = \text{tr } T = k(T_d).$$

It is easily checked that the mapping $T \to \det T$ of $\mathfrak{gl}(V)$ into F is a homomorphism on $Gl(V) \subset \mathfrak{gl}(V)$, and that $T \in Gl(V)$ if and only if $\det T \neq 0$. It is also easy to show that this definition of determinant coincides with the more usual ones.

Problem 8. Show (a) $\text{tr}(S + T) = \text{tr}(S) + \text{tr}(T)$
(b) $\text{tr}(ST) = \text{tr}(TS)$. (Use problem 7 to do this without using coordinates.)

Problem 9. Let f be in the dual space of $\mathfrak{gl}(V)$ and satisfy: (a) $f(ST) = f(TS)$ for every $T, S \in \mathfrak{gl}(V)$, (b) $f(\text{identity}) = d = \text{dimension } V$. Prove that $f = \text{tr}$ and that $[\mathfrak{gl}(V), \mathfrak{gl}(V)]$ has co-dimension 1 in $\mathfrak{gl}(V)$ with a complementary space consisting of multiples of $I = \text{identity}$.

4.4 Existence of Grassmann Algebras

Alternating Functions. The particular Grassmann algebra over V which we wish to consider will consist of the space of multilinear alternating functions on the dual W of V. Thus we will have

$$G^0 = F, \text{ the field of } V,$$

$$G^1 = \text{linear functions on } W \text{ into } F.$$

G^1 is then canonically isomorphic to V in the usual way: $v \in V$ corresponds to the unique $f \in G^1$ such that $f(w) = w(v)$ for all $w \in W$. This isomorphism will be our imbedding of V in G.

$G^p =$ the linear space of p-linear alternating functions from $W \times \cdots \times W$ (p factors) into F. More explicitly, if $f \in G^p$, then

(i) $\quad f(w_1, ..., aw_i + bw_i', ..., w_p) = af(w_1, ..., w_i, ..., w_p)$

$$+ bf(w_1, ..., w_i', ..., w_p),$$

that is, f is linear in each variable when the others are fixed,

(ii) $f(w_1, ..., w_p) = 0$

whenever two of the w_i's are equal.

If the characteristic of F is not 2, then we can use the standard polarization method to show that (i) and (ii) are equivalent to (i) and

(ii)′ $f(w_1, ..., w_i, ..., w_j, ..., w_p)$
$$= -f(w_1, ..., w_j, ..., w_i, ..., w_p).$$

Moreover, since transpositions generate the symmetric group S_p, (ii)′ is equivalent to

(ii)″ $f(w_{\pi 1}, ..., w_{\pi p}) = \operatorname{sgn}(\pi) f(w_1, ..., w_p)$

for every $\pi \in S_p$. Even for characteristic 2, (i) and (ii) imply (ii)′.

It will sometimes be convenient to regard $W \times \cdots \times W$ (p factors) as functions from $\{1, \cdots, p\}$ into W, and then we would rewrite (ii)″ as

$$f(z \circ \pi) = \operatorname{sgn}(\pi) f(z) \qquad \text{for} \qquad z \in W \times \cdots \times W.$$

Multiplication. Let S_{p+q} be the permutation group on $\{1, ..., p+q\}$, S_p the subgroup leaving $\{p+1, ..., p+q\}$ pointwise fixed, S_q' the subgroup leaving $\{1, ..., p\}$ pointwise fixed. By a *cross section of* $S_p S_q'$ *in* S_{p+q} we mean a subset K of S_{p+q} having one and only one element from each left coset of $S_p S_q'$. A particular cross section which is often used is the set of *shuffle permutations*, namely,

$$K = \{\pi \in S_{p+q} \mid \pi 1 < \pi 2 < \cdots < \pi p \qquad \text{and}$$
$$\pi(p+1) < \pi(p+2) < \cdots < \pi(p+q)\}.$$

The name is derived from the fact that if a deck of $p+q$ cards is cut into two stacks of p and q cards and then shuffled together with the usual technique, then the resulting permutation on the deck is a p, q shuffle permutation.

FIG. 15.

If $f \in G^p$, $g \in G^q$, K a cross section of $S_p S_q{}'$ in S_{p+q}, we let A_p be the operation of adding p, and then for $z \in W^{p+q}$ we define the product fg by

$$fg(z) = \sum_{\pi \in K} \text{sgn}\,(\pi)\, f(z \circ \pi)\, g(z \circ \pi \circ A_p).$$

To clarify notation, if $z(i) = w_i$ note that

$$f(z \circ \pi) = f(w_{\pi 1}, ..., w_{\pi p})$$

$$g(z \circ \pi \circ A_p) = g(w_{\pi(p+1)}, ..., w_{\pi(p+q)}).$$

Using (ii)″ it is easy to show that this multiplication is independent of K.

Proposition 1. $fg \in G^{p+q}$.

Proof. $(p + q)$-linearity of fg is obvious. Suppose $\sigma \in S_{p+q}$, K a cross section as above. Then

$$fg(z \circ \sigma) = \sum_{\pi \in K} \text{sgn}\,(\pi)\, f(z \circ \sigma \circ \pi)\, g(z \circ \sigma \circ \pi \circ A_p)$$

$$= \text{sgn}\,(\sigma) \sum_{\rho \in \sigma K} \text{sgn}\,(\rho)\, f(z \circ \rho)\, g(z \circ \rho \circ A_p).$$

Then it suffices to show that σK is a cross section also. We have if π, $\pi' \in K$, $(\sigma\pi)^{-1}(\sigma\pi') = \pi^{-1}\pi'$ which is not in $S_p S_q{}'$ unless $\pi = \pi'$. Thus the elements of σK are all in different cosets, and since the cardinality of this set is right, it must be a cross section.

Notice that we have only shown that (ii)′ holds for fg. The proposition is true for characteristic 2, but we do not prove it.

Proposition 2. This multiplication is associative.

Proof. (Outline). Let $f \in G^p$, $g \in G^q$, $h \in G^r$. We consider S_{p+q} as a subgroup of S_{p+q+r}, and $S_r{}''$ the permutations on $\{p + q + 1, ..., p + q + r\}$ and similar conventions for S_p and $S_{q+r}{}'$. Then we show that if K is a cross section of $S_p S_q{}'$ in S_{p+q}, K' of $S_{p+q} S_r{}''$ in S_{p+q+r}, L of $S_q{}' S_r{}''$ in $S_{q+r}{}'$, and L' of $S_p S_{q+r}{}'$ in S_{p+q+r}, then $K'K$ and $L'L$ are both cross sections of $S_p S_q{}' S_r{}''$ in S_{p+q+r}. Thus if we write out the products associated in both ways we see that we need only show that a sum

$$fgh(z) = \sum_{\pi \in K'K} \text{sgn}\,(\pi)\, f(z \circ \pi)\, g(z \circ \pi \circ A_p)\, h(z \circ \pi \circ A_{p+q})$$

is independent of the cross section of $S_p S_q' S_r''$ in S_{p+q+r} over which we sum. This again is trivial by (ii)''.

Proposition 3. For $v_1, ..., v_p \in V$ we have

$$v_1 \cdots v_p(w_1, ..., w_p) = \sum_{\pi \in S_p} \text{sgn} (\pi) \, v_1(w_{\pi 1}) \cdots v_p(w_{\pi p}).$$

The proof is just an extension of the arguments for the proof of proposition 2.

Corollary. If $v \in V$, then $v^2 = 0$.

Finally, addition, scalar multiplication, and the identity are defined in the obvious ways.

Dimension and Generation by $\{e, V\}$. We have already shown that $G = \Sigma G^i$ satisfies axioms (i)–(iii) for a Grassmann algebra. It remains to prove (iv) and (v).

Lemma 2. Let $e_1, ..., e_d$ be a basis for V. Then $e_D \neq 0$, where $D = \{1, ..., d\}$.

Proof. Let $w_1, ..., w_d$ be a dual basis to $e_1, ..., e_d$. Then by proposition 3 above $e_D(w_1, ..., w_d) = 1 \neq 0$.

Proposition 4. The e_s's are linearly independent, so dim $G \geqslant 2^d$.

Proof. Since the sum $\Sigma \, G^i$ is direct we need only show that the e_s's, with s all of cardinality p, are independent. Let P be the set of all such s. Now if we have $\Sigma_{s \in P} \, a_s e_s = 0$ and if $s_0 \in P$, then

$$0 = \sum_{S \in P} a_s e_s e_{D-s_0} = \pm \, a_{s_0} e_D \,,$$

since $e_i^2 = 0$ and the e_i anticommute. Hence, by the lemma 2, $a_{s_0} = 0$. QED

Proposition 5. The e_s's span G, so dim $G = 2^d$.

Proof. Let $w_1, ..., w_d$ be a basis for W. For $s = \{i_1, ..., i_p\}$, $i_1 < i_2 < \cdots < i_p$, let $w_s = (w_{i_1}, ..., w_{i_p})$. Let $e_1, ..., e_d$ be the dual basis for V. Then by proposition 3, $e_s(w_s) = 1$, $e_s(w_t) = 0$ if $s \neq t$.

Now for $f \in G^p$ we have

$$f\left(\sum a_{1i}w_i, ..., \sum a_{pi}w_i\right) = \sum_{i_1,...,i_p} a_{1i_1} \cdots a_{pi_p} \, f(w_{i_1}, ..., w_{i_p})$$

$$= \sum_{s \in P} \sum_{\{i_1,...,i_p\}=s} \operatorname{sgn}\,(\pi(i_1, ..., i_p))\, a_{1i_1} \cdots a_{pi_p} f(w_s),$$

where $\pi(i_1, ..., i_p)$ is the permutation needed to put $i_1, ..., i_p$ in ascending order.

Applying this formula with $f = e_s$ gives

$$e_s\left(\sum a_{1i}w_i, ..., \sum a_{pi}w_i\right) = \sum_{\{i_1,...,i_p\}=s} \operatorname{sgn}\,(\pi(i_1, ..., i_p))\, a_{1i_1} \cdots a_{pi_p}.$$

Hence $f = \Sigma_{s \in P} f(w_s)\, e_s$.

Problem 10. Prove that the shuffle permutations are a cross section and use them to write out all possible products for $p, q \leqslant 2$.

Problem 11. Suppose we define another product on the collection of multilinear alternating functions on W by: if $f \in G^p$, $g \in G^q$, $p, q \geqslant 1$, then $f*g = a_{pq}fg$, where a_{pq} is a nonzero element of F, fg is the old product. In order that $G = \Sigma G^i$ be a Grassmann algebra over V with $*$ as multiplication we must have $a_{pq} = a_{qp}$ and a condition equivalent to the associative law. Derive this condition and use it to express a_{pq} in terms of $a_{11}, a_{12}, ..., a_{1d-1}$. Conversely, show that a_{1j} may be assigned arbitrarily, and that when we let $a_{1j} = j + 1$ the formula for $f*g$ is

$$f*g(z) = \sum_{\pi \in S_{p+q}} \operatorname{sgn}\,(\pi)\, f(z \circ \pi)\, g(z \circ \pi \circ A_p).$$

This formula may be used to define multiplication instead of the one we have given, but it does not work when the characteristic of F is a prime $\leqslant d$. Our formula has the fewest possible terms.

Problem 12. Let $e = \{e_1, ..., e_d\}$ be a basis of V, $\mathscr{L} = \mathfrak{gl}(V)$, $J: \mathscr{L} \to V^d$ the isomorphism of $\mathscr{L}$ onto the d-fold Cartesian product of V given by $J(T) = (Te_1, ..., Te_d)$, $f \in G^d(W)$ the unique element of $G^d(W)$ such that $f(e_1, ..., e_d) = 1$ (f is an alternating d-linear function on V^d). Prove that $\det = f \circ J: \mathscr{L} \to F$.

Problem 13. Let $f \in G^d(F^{d*})$ be the unique alternating d-linear function on $(F^d)^d = d \times d$ matrices such that $f(I) = 1$. Prove that f is the usual determinant of matrices.

Problem 14. Describe a natural isomorphism of $G(V^*)$ and $G(V)^*$. Use this to show that an inner product on V extends naturally to an inner product on $G(V)$, since an inner product leads to an isomorphism of V and V^* which extends uniquely to an algebra isomorphism of $G(V)$ and $G(V^*)$. Also work out the expressions for the inner product on $G(V)$ in terms of a basis.

4.5 Differential Forms

Henceforth we will be concerned with the case $V = M_m{}^*$, $m \in M$, a manifold, $W = M_m$, $G_m = G(M_m{}^*)$. Thus if $(x_1, ..., x_d)$ is a coordinate system at m, $dx_1, ..., dx_d$ at m form a basis of $M_m{}^*$, so that every element f of $G_m{}^p$ is uniquely expressible as $f = \Sigma_{s \in P} a_s \, dx_s$.

A *differential p-form* of M is a function θ defined on some subset E of M, whose value at each $m \in E$ is an element of $G_m{}^d$. θ is a C^∞ *p-form* if, for each set of C^∞ vector fields $V_1, ..., V_p$ on M, the function $\theta(V_1, ..., V_p)$ defined on the intersection of the domains of θ and of the $V_1, ..., V_p$ by

$$\theta(V_1, ..., V_p)(m) = \theta_m(V_1(m), ..., V_p(m))$$

is C^∞. Here θ_m denotes the value of θ at m. When there is no confusion the "m" will be dropped.

A 0-form is simply a real valued function on M. We notice that θ is C^∞ if and only if for each coordinate system $(x_1, ..., x_d)$ and (unique) expression $\theta = \Sigma_{s \in P} a_s \, dx_s$ we have $a_s \in C^\infty$.

Alternative Approach. We now relate our definition of forms to the brief discussion of tensors given in 3.3. First form the following tensor bundles over M:

$$G^p = \bigcup_{m \in M} G_m{}^p.$$

The projection is the natural one, $\pi : G_m{}^p \to m$. The differentiable structure is given by taking as coordinate neighborhoods sets of the form $\bar{V} = \bigcup_{m \in V} G_m{}^p$, where V is a coordinate neighborhood in M with coordinates $x_1, ..., x_d$, and $\bar{V}$ has coordinates $x_1 \circ \pi, ..., x_d \circ \pi$

4.5. Differential Forms

plus the $\binom{d}{p}$ functions w_s, $s \in P$, dual to the basis $\{dx_s\}$ of $G_m{}^p$ at each point of V.

Alternatively, if R^{d*} is the dual of Euclidean d-space R^d, then by dualizing the action of $Gl(d, R)$ on R^d we see that $Gl(d, R)$ acts on R^{d*} by taking the transpose of the inverse and hence also on $G(R^{d*})^p$ by extending to homomorphisms. G^p is then the bundle associated to the bundle of bases $B(M)$ by this action [see 3.3(3)].

A *differential p-form* θ on $E \subset M$ is then a cross section of G^p over E, that is, a map $\theta : E \to G^p$ such that $\pi \circ \theta =$ identity. $\theta \in C^\infty$ if θ is a C^∞ map.

Orientation. If E is a vector bundle over M we let $E_0 =$ the set of nonzero vectors of E. E_0 is an open submanifold of E.

If M is connected then $G^d{}_0$ is either connected or has exactly two components. This follows easily from the facts that any path in M can be lifted to a path in $G^d{}_0$ and each fibre of $G^d{}_0$ has exactly two components ($d =$ dimension of M).

M is *orientable* if $G^d{}_0$ has two components. Each component of $G^d{}_0$ is in this case called an orientation of M. M is *nonorientable* if $G^d{}_0$ is connected.

Lemma 3. A paracompact manifold M is orientable if and only if M admits a continuous, nonvanishing, globally defined d-form.

Proof. Suppose θ is a continuous, everywhere defined d-form with values in $G^d{}_0$. Let γ be any curve in $G^d{}_0$ which begins and ends in the same fibre $G_m{}^d{}_0 =$ nonzero elements of $G_m{}^d$. Since $G_n{}^d$ is one-dimensional for each $n \in M$, we may write $\gamma(t) = \alpha(t) \theta(\pi \circ \gamma(t))$, where α is a continuous real-valued function. Now $\alpha(t)$ is never zero, so $\alpha(t)$ is of constant sign. Hence, the initial and final points of γ are in the same component of $G_m{}^d{}_0$. Thus there is no curve connecting the two components of $G_m{}^d{}_0$, so $G^d{}_0$ is not connected and M is orientable.

Conversely, suppose that M is orientable. Let $\{U_i\}$ be a locally finite covering of M by connected coordinate neighborhoods, $\{f_i\}$ an associated C^∞ partition of unity. Choose an orientation of M. Then for any connected coordinate neighborhood U with associated coordinates $x_1, ..., x_d$, $\phi = dx_D$ is a continuous cross section of U in $G^d{}_0$, so either ϕ or $-\phi$ is in the chosen orientation. Thus we get ϕ_i defined on U_i with values in the chosen orientation and $\theta = \Sigma f_i\phi_i$ defines the desired nonzero C^∞ d-form. QED

For an orientable M such a θ is called a *volume element of M*.

If M is not connected, then we say that M is orientable if every component of M is orientable, and an *orientation* of M is a choice of orientations for every component.

There are other conditions equivalent to orientability which are sometimes used as the definition. They are

(1) M is covered by coordinate neighborhoods in such a way that any two systems are related by a system of equations having positive Jacobian determinant.

(2) (When M is connected.) The bundle of bases $B(M)$ is not connected. $B(M)$ will then have just two components, each representing an orientation of M.

Problem 15. Prove that the following manifolds are orientable:

(a) The tangent bundle of any manifold.

(b) A parallelizable manifold, hence, Lie groups.

(c) Complex manifolds and almost complex manifolds.

4.6 Exterior Derivative

The *exterior derivative d* is a map which assigns to each C^∞ p-form θ a C^∞ $(p+1)$-form $d\theta$ such that

(i) if $p = 0$, $d\theta$ agrees with the definition of the differential of a C^∞ function (see 4.1),

(ii) domain of θ = domain of $d\theta$,

(iii) d is R-linear,

(iv) if θ is a p-form, ϕ a q-form, then $d(\theta\phi) = (d\theta)\phi + (-1)^p\theta(d\phi)$,

(v) $d(d\theta) = 0$ for all θ.

Theorem 1. There is at most one map d satisfying (i)-(v) above.

Proof. If we write $\theta = \Sigma_{s\in P}\, a_s\, dx_s$ on the intersection of the domains of θ and the x_i, we must have

$$d\theta = \sum_{s\in P} d(a_s\, dx_s) \qquad \text{[by (iii)]}$$

$$= \sum_{s\in P} (da_s\, dx_s + a_s\, d(dx_s)) \qquad \text{[by (iv)].} \qquad (1)$$

Now by (iv) and (v)

$$d(dx_s) = \sum_{j=1}^{p} (-1)^{j-1} \, dx_{i_1} \cdots d(dx_{i_j}) \cdots dx_{i_p}$$

$$= 0.$$

Thus we get $d\theta = \Sigma_{s \in P} \Sigma_{j=1}^{p} D_{x_j} a_s \, dx_j \, dx_s$ as the unique expression for $d\theta$.

There is one point that has been passed over; namely, when we wrote (1), we should actually have put for the left-hand side

$$d(\theta \mid_{(\text{domain } \theta \, \cap \, \text{domain } \{x_i\})}),$$

and what we claim is that this coincides with

$$d\theta \mid_{(\text{domain } \theta \, \cap \, \text{domain } \{x_i\})}.$$

A similar remark holds for the right-hand side of (1). Thus we still must show that if θ and ϕ agree on an open set U, then $d\theta$ and $d\phi$ also agree on U. To do this we let f be the function with domain U and value 1 at every point of U. Then $f\theta$ and $f\phi$ are equal and have the same domain U. By (i) $df = 0$, so by (iv)

$$d\theta \mid_U = f \, d\theta = d(f\theta) = d(f\phi) = f \, d\phi = d\phi \mid_U,$$

as desired. This concludes the proof of the theorem.

Thus we have demonstrated the form which d must take in a coordinate system. In order to show that d exists we would have to show consistency in overlapping coordinate systems. This is actually quite easy, since we can verify (i)-(v) for a coordinate neighborhood, and then the uniqueness and consistency under restriction to smaller domain will give the consistency on the intersections of coordinate neighborhoods. Now we develop an intrinsic formulation of d which will be used extensively later.

Intrinsic Formula for d. Let θ be a C^∞ p-form. Then we define an R-$(p + 1)$-linear function $\bar\theta$ of $(p + 1)$-tuples of C^∞ vector fields into C^∞ functions on M by

$$\bar\theta(V_1, ..., V_{p+1}) = \sum_i (-1)^{i-1} V_i \theta(V_1, ..., V_{i-1}, V_{i+1}, ..., V_{p+1})$$

$$+ \sum_{i<j} (-1)^{i+j} \theta([V_i, V_j], V_1, ..., V_{i-1}, V_{i+1}, ..., V_{j-1}, V_{j+1}, ..., V_{p+1}).$$

If $t_1, ..., t_{p+1} \in M_m$, $m \in$ domain of θ, choose vector fields $V_1, ...,$ V_{p+1} such that $V_i(m) = t_i$. Define $d\theta(t_1, ..., t_{p+1}) = \bar{\theta}(V_1, ..., V_{p+1})(m)$. We must show that this definition is independent of the way in which $V_1, ..., V_{p+1}$ are chosen. This will be proved in a series of lemmas.

Lemma 4. If V_i and W_i coincide in a neighborhood of m for $i = 1, ..., p + 1$, then $\bar{\theta}(V_1, ..., V_{p+1})(m) = \bar{\theta}(W_1, ..., W_{p+1})(m)$.

Proof.   This is clear from the definition of $\bar{\theta}$.

Lemma 5. (a) $\bar{\theta}(fV_1, V_2, ..., V_{p+1}) = f\bar{\theta}(V_1, ..., V_{p+1})$.
(b) $\bar{\theta}$ is alternating, that is,

$$\bar{\theta}(V_1, ..., V_i, ..., V_j, ..., V_{p+1}) = -\bar{\theta}(V_1, ..., V_j, ..., V_i, ..., V_{p+1}).$$

(c) $\bar{\theta}(V_1, ..., fV_i, ..., V_{p+1}) = f\bar{\theta}(V_1, ..., V_{p+1})$.

Proof. (a) follows from problem 1.14 and the definition of $\bar{\theta}$. (b) follows immediately from the definition of $\bar{\theta}$. (c) then follows from (a) and (b).

Lemma 6. $d\theta$ is well defined, that is, $\bar{\theta}(V_1, ..., V_{p+1})(m)$ is independent of the choice of V_i such that $t_i = V_i(m)$.

Proof.   Let $x_1, ..., x_d$ be coordinates at m. Then $V_i = \Sigma_j a_{ij} D_{x_j}$ in a neighborhood of m. If $W_i(m) = V_i(m)$, then $W_i = \Sigma_j b_{ij} D_{x_j}$ and $a_{ij}(m) = b_{ij}(m)$ for all i, j. By lemma 4 we need only to consider $\bar{\theta}$ in this coordinate system. By (c) of lemma 5,

$$\bar{\theta}(V_1, ..., V_{p+1})(m) = \sum_{i=1}^{p+1} \sum_{j_i=1}^{d} a_{ij_i}(m) \, \bar{\theta}(D_{x_{j_1}}, ..., D_{x_{j_{p+1}}})(m)$$

$$= \bar{\theta}(W_1, ..., W_{p+1})(m). \qquad \text{QED}$$

Theorem 2.   The map $\theta \to d\theta$ defined via $\bar{\theta}$ above is the exterior derivative, that is, satisfies properties (i)-(v).

Proof.   Properties (ii) and (iii) are obvious from the definition. We verify property (i) directly; namely, let $\theta = f \in C^\infty$, and we show $d\theta = df$. Let $t \in M_m$, V a vector field such that $V(m) = t$. But $\bar{\theta}(V) = (-1)^{1-1} Vf$, the other terms being vacuous. Hence, at m

$$d\theta(t) = \bar{\theta}(V)(m) = V(m)f = tf = df(t), \qquad \text{as desired.}$$

To show the other properties we establish that d coincides with the operation defined locally in the proof of theorem 1. Let $\tilde{d}$ denote this operation, that is, if $x_1, \ldots, x_d$ are coordinates, then for

$$\theta = \sum_{s \in P} a_s \, dx_s$$

$$\tilde{d}\theta = \sum_{s \in P} \sum_j D_{x_j} a_s \, dx_j \, dx_s.$$

Now the assignments $\theta \to d\theta$ and $\theta \to \tilde{d}\theta$ are both R-linear. Further, if θ and ϕ agree on a neighborhood then $d\theta$ and $d\phi$ also agree on this neighborhood. Hence it is sufficient to show they are the same for $\theta = f \, dx_s$, and by linearity of forms it is enough to show that $d\theta$ and $\tilde{d}\theta$ agree when applied to $(V_1, \ldots, V_{p+1})$, where $V_i = D_{x_{j_i}}$:

$$\tilde{d}\theta(V_1, \ldots, V_{p+1}) = \sum_k D_{x_k} f \, dx_k \, dx_s(V_1, \ldots, V_{p+1})$$

$$= \sum_k D_{x_k} f \sum_{\pi \in K} \operatorname{sgn}(\pi) \, dx_k(V_{\pi 1}) \, dx_s(V_{\pi 2}, \ldots, V_{\pi(p+1)}),$$

where K is a cross section of $S_1 S_p$ in S_{p+1}. Using shuffle permutations $\pi 2 < \cdots < \pi(p + 1)$ we get

$$= \sum_k D_{x_k} f \sum_{m=1}^{p+1} (-1)^{m-1} \delta_{k j_m} \, dx_s(V_1, \ldots, V_{m-1}, V_{m+1}, \ldots, V_{p+1})$$

$$= \sum_m (-1)^{m-1} D_{x_{j_m}} f \, dx_s(V_1, \ldots, V_{m-1}, V_{m+1}, \ldots, V_{p+1}).$$

On the other hand, since $[V_i, V_j] = 0$,

$$d\theta(V_1, \ldots, V_{p+1}) = \sum_{m=1}^{p+1} (-1)^{m-1} V_m(f \, dx_s(V_1, \ldots, V_{m-1}, V_{m+1}, \ldots, V_{p+1})),$$

and since $dx_s(V_1, \ldots, V_{m-1}, V_{m+1}, \cdots, V_{p+1}) = $ constant,

$$= \sum_m (-1)^{m-1} D_{x_{j_m}} f \, dx_s(V_1, \ldots, V_{m-1}, V_{m+1}, \ldots, V_{p+1}). \qquad \text{QED}$$

Problem 16. Write out in full the intrinsic expression for $d\theta$ when θ is a 1-form and a 2-form. In the latter case use skew-symmetry to put it in the form of a sum over the cyclic permutations of 1, 2, 3.

Problem 17. If U, V, W are constant vector fields (brackets all 0) on R^3 and X is a C^∞ vector field, set up correspondences between the various parts of the calculus of differential forms and ordinary vector analysis to obtain the following formulas:

from the intrinsic formula for d:

$$\operatorname{grad} f \cdot U = Uf, \qquad \operatorname{curl} X \cdot U \times V = U(X \cdot V) - V(X \cdot U)$$

$$(\operatorname{div} X) \, U \cdot V \times W = U(X \cdot V \times W) + V(X \cdot W \times U) + W(X \cdot U \times V)$$

(These formulas could be used to define grad, curl, and div.)

from axiom (iv) for d:

$$\operatorname{grad}(fg) = g \operatorname{grad} f + f \operatorname{grad} g$$
$$\operatorname{curl}(fX) = (\operatorname{grad} f) \times X + f \operatorname{curl} X$$
$$\operatorname{div}(fX) = (\operatorname{grad} f) \cdot X + f \operatorname{div} X$$
$$\operatorname{div}(X \times Y) = (\operatorname{curl} X) \cdot Y - X \cdot \operatorname{curl} Y$$

from axiom (v) for d:

$$\operatorname{curl} \operatorname{grad} f = 0, \qquad \operatorname{div} \operatorname{curl} X = 0.$$

4.7 Action of Maps

If M, N are manifolds, $\phi : M \to N$ a C^∞ map, then we have seen that ϕ carries functions on N into functions on M by composition and carries tangents on M into tangents on N. We now define a map ϕ^* of forms on N into forms on M, namely, $\phi^* : \theta \to \theta \circ d\phi$, that is, if θ is a p-form on N, then

$$\phi^*(\theta)(t_1, \dots, t_p) = \theta(d\phi \, t_1, \dots, d\phi \, t_p).$$

Notice that on the space of 1-forms at a point ϕ^* is just the usual dual linear transformation to $d\phi$.

Theorem 3. ϕ^* is a Grassmann algebra homomorphism and it commutes with the exterior derivative.

Proof. That ϕ^* is a Grassmann algebra homomorphism is automatic from the definition of multiplication.

To prove that ϕ^* commutes with d we notice first that both ϕ^* and d are R-linear and that it is a local problem. Locally forms are generated by functions and differentials of functions. Using the fact that ϕ^* is a homomorphism and d is an antiderivation the problem is reduced to consideration of individual factors of terms, namely, to functions and differentials of functions. But we have for any function f on N

$$\phi^* \, df = df \circ d\phi = d(f \circ \phi) = d(\phi^* f) \quad \text{and}$$
$$d(\phi^* \, df) = d(d(f \circ \phi)) = 0 = \phi^*(0) = \phi^*(d(df)). \quad \text{QED}$$

Example. If $\theta = u_2 \, du_1 + u_3 \, du_2 + u_1 \, du_3$ and $\phi : R^2 \to R^3$ by $\phi = (\sin u_1 \cos u_2 \, , \, \sin u_1 \sin u_2 \, , \, \cos u_1)$, then

$$\phi^* \theta = \sin u_1 \sin u_2 \, d(\sin u_1 \cos u_2) + \cos u_1 \, d(\sin u_1 \sin u_2)$$
$$+ \sin u_1 \cos u_2 \, d \cos u_1$$
$$= (\sin u_1 \cos u_1 \sin u_2 \cos u_2 + \cos^2 u_1 \sin u_2 - \sin^2 u_1 \cos u_2) \, du_1$$
$$+ (-\sin^2 u_1 \sin^2 u_2 + \sin u_1 \cos u_1 \cos u_2) \, du_2.$$

Problem 18. Denote by $\mathcal{K}(M)$ the collection of all C^∞ forms defined on open subsets of M, $\mathcal{K}^p(M)$ the collection of C^∞ p-forms; sometimes we shall omit "(M)."

Let X be a vector field and define a linear function $i(X) : \mathcal{K} \to \mathcal{K}$ satisfying the following conditions:

(a) $i(X) : \mathcal{K}^p \to \mathcal{K}^{p-1}$, $p \geqslant 1$, $i(X)(\mathcal{K}^0) = 0$.

(b) $i(X)$ is an antiderivation, that is, if $\theta \in \mathcal{K}^p$, $\phi \in \mathcal{K}^q$, then $i(X)(\theta\phi) = (i(X) \, \theta) \, \phi + (-1)^p \theta(i(X) \, \phi)$.

(c) If $\theta \in \mathcal{K}^1$, then $i(X) \, \theta = \theta(X)$.

Show that there is a unique function satisfying (a)-(c), and verify the formula $i(X) \, \theta(X_1, ..., X_{p-1}) = \theta(X, X_1, ..., X_{p-1})$, where $\theta \in \mathcal{K}^p$, and $X_1, ..., X_{p-1}$ are vector fields.

Problem 19. Show that L_X, the Lie derivative of X, restricted to $\mathcal{K}$ is a derivation. Show that $i(X) \, d + di(X)$ is a derivation. Hence prove the formula

$$L_X = i(X) \, d + di(X)$$

by showing that it holds for functions and differentials of functions.

Problem 20. Let G be a Lie group, $X_1, ..., X_d$ a basis for the left invariant vector fields on G, $c_{ij}{}^k$ the constants (*structural constants*) such that $[X_i, X_j] = \Sigma_k c_{ij}{}^k X_k$. Define dual 1-forms $\omega_1, ..., \omega_d$ to the X_i by $\omega_i(X_j) = \delta_{ij}$ (constant function). Prove

(a) ω_i is left invariant, that is, for every $g \in G$, $L_g{}^* \omega_i = \omega_i$.

(b) $d\omega_i = -\frac{1}{2}\Sigma_{j,k} c_{jk}{}^i \omega_j \omega_k = \Sigma_{j<k} c_{jk}{}^i \omega_k \omega_j$. (This is called the *equation of Maurer-Cartan.*)

4.8 Frobenius' Theorem.

A *k-dimensional codistribution* E on a manifold M is a map which assigns to each $m \in M$ a k-dimensional subspace E_m of $G_m{}^1$. E is a C^∞ *codistribution* if for each $m \in M$ there is a neighborhood on which are defined k C^∞ 1-forms which span E at each point of the neighborhood. A submanifold N of M is an *integral submanifold of E* if for each $n \in N$, E_n = annihilator of $dI(N_n)$, where $I: N \to M$ is the injection map. E is *completely integrable* if there is an integral submanifold of E through each point $m \in M$. The *distribution associated with E* is the $(d - k)$-distribution D defined by D_m = annihilator of E_m.

Lemma 7. Let W_m be the ideal generated by E_m in G_m. Then $\omega \in W_m{}^p$ if and only if for every $t_1, ..., t_p \in D_m$ we have $\omega(t_1, ..., t_p) = 0$.

Proof. If $\theta_1, ..., \theta_k$ is a basis for E_m, then if $\omega \in W_m{}^p$ we have $\omega = \Sigma_i \phi_i \theta_i$, $\phi_i \in G_m{}^{p-1}$. Thus

$$\omega(t_1, ..., t_p) = \sum_i \phi_i \theta_i(t_1, ..., t_p)$$

$$= \sum_{i,j} (-1)^{p-j} \phi_i(t_1, ..., t_{j-1}, t_{j+1}, ..., t_p)\, \theta_i(t_j)$$

$$= 0$$

since $t_j \in D_m$ = annihilator of E_m.

Conversely, if $\omega(t_1, ..., t_p) = 0$ for every $t_1, ..., t_p \in D_m$, let $\theta_{k+1}, ..., \theta_d$ be a completion of $\theta_1, ..., \theta_k$ to a basis of $G_m{}^1$, and let $e_1, ..., e_d$ be a basis dual to $\theta_1, ..., \theta_d$. Then $e_{k+1}, ..., e_d$ span D_m, and if we write $\omega = \Sigma_{s \in P} a_s \theta_s$, we have, if $k < j_1 < \cdots < j_p$,

$$0 = \omega(e_{j_1}, ..., e_{j_p}) = \sum_{s \in P} a_s \theta_s(e_{j_1}, ..., e_{j_p})$$

$$= a_{j_1 ... j_p}.$$

Hence each term of ω must have a θ_i with $i \leqslant k$, that is, $\omega \in W_m{}^p$.

Frobenius' Theorem [29]. Let E be a C^∞ k-dimensional codistribution on M. Let W be the function which assigns to each $m \in M$ the ideal W_m generated by E_m in G_m. Then E is completely integrable if and only if W is invariant under the exterior derivative operator, that is, $d(W) \subset W$.

We shall show that this theorem is equivalent to theorem 1.6, which says that a C^∞ distribution is completely integrable if and only if it is involutive. First notice that the condition $d(W) \subset W$ is equivalent to the assertion that for each $m \in M$ there is a local basis of 1-forms θ_1, ..., θ_k in E such that $d\theta_i \in W$.

We have E is completely integrable if and only if the associated $(d - k)$-distribution D is completely integrable if and only if D is involutive if and only if $\theta_i(V_j) = 0$, $i = 1$, ..., k; $j = 1$, 2 implies $\theta_i([V_1, V_2]) = 0$, $i = 1$, ..., k if and only if $\theta_i(V_j) = 0$, $i = 1$, ..., k; $j = 1$, 2 implies $d\theta_i(V_1, V_2) = 0$, $i = 1$, ..., k, (theorem 2) if and only if $d\theta_i \in W$, $i = 1$, ..., k (lemma 7). QED

Problem 21. Derive the integrability condition given in problem 1.29 for the 1-codistribution spanned by the 1-form $P\,du_1 + Q\,du_2 + R\,du_3$. (*Hint*: Use problem 4.)

Problem 22. Let ω be a p-form with $p \leqslant d - k$. Prove that ω is in the ideal generated by linearly independent 1-forms θ_1, ..., θ_k if and only if $\omega\theta_1 \ldots \theta_k = 0$.

On the other hand, if $p > d - k$, then ω is always in the ideal generated by θ_1, ..., θ_k, so this shows: W is the space annihilated by multiplication by $\theta_1 \ldots \theta_k$.

4.9 Vector-Valued Forms and Operations

Let V be a vector space over R, M a manifold. A *V-valued p-form on M* is a map ω which assigns to each t_1, ..., $t_p \in M_m$ an element $\omega(t_1, ..., t_p)$ of V such that if f is any element of the dual space of V then $f \circ \omega$ is a (*real-valued*) p-form on M.

Examples. (1) If $V = R^d$, then a V-valued p-form is simply a d-tuple of p-forms $(\omega_1, ..., \omega_d) = (u_1 \circ \omega, ..., u_d \circ \omega)$.

(2) If $V = \mathfrak{gl}(R^d)$, the Lie algebra of $d \times d$ matrices, then a V-valued p-form is just a matrix of p-forms (ω_{ij}).

If ϕ is a $\mathfrak{g}$-valued p-form, ω a $\mathfrak{g}$-valued q-form, where $\mathfrak{g}$ is a Lie algebra, then $[\phi, \omega]$ is the $\mathfrak{g}$-valued $(p + q)$-form given by

$$[\phi, \omega](t) = \sum_{\pi \in K} \operatorname{sgn}(\pi)[\phi(t \circ \pi), \omega(t \circ \pi \circ A_p)],$$

where K is a cross section of $S_p S_q'$ in S_{p+q}.

Note that $[\phi, \omega] = (-1)^{pq-1}[\omega, \phi]$ since we get an additional factor of -1 when we change the order of bracketing, besides the factor $(-1)^{pq}$ resulting from reversing the order of the forms.

If ϕ is a p-form with values in a space of linear transformations on a vector space V, and ω is a V-valued q-form, then $\phi\omega$ is defined as a V-valued $(p + q)$-form by

$$\phi\omega(t) = \sum_{\pi \in K} \operatorname{sgn}(\pi)\, \phi(t \circ \pi)\, \omega(t \circ \pi \circ A_p),$$

where K is again a cross section as above.

Problem 23. Let G be a Lie group, $\mathfrak{g}$ its Lie algebra. Define $\mathfrak{g}$-valued 1-form ω on G by $\omega(X) = X$ for every $X \in \mathfrak{g}$, that is, if $x = X(g)$ then $\omega(x) = X$.

   (a) Prove that ω is left invariant.
   (b) Prove that $d\omega = -\frac{1}{2}[\omega, \omega]$.
   (c) If X_1, ..., X_d is a basis for $\mathfrak{g}$, then we may write $\omega(x) = \Sigma_i \omega_i(x) X_i$, defining real-valued 1-forms ω_1, ..., ω_d. Prove that ω_1, ..., ω_d are the same as in problem 20, so that (b) is a coordinate-free form of the equations of Maurer-Cartan.

Problem 24. (a) If ϕ, ω are $\mathfrak{gl}(d, R)$-valued forms, define the product $\phi\omega$ using matrix multiplication instead of bracket multiplication as above.

   (b) Show that $[\phi, \phi](X, Y) = 2[\phi(X), \phi(Y)] = 2\phi\phi(X, Y)$ when ϕ is a 1-form.

4.10 Forms on Complex Manifolds [23, 30, 50, 92]

If M is a complex manifold, then the differential forms are defined in terms of Grassmann algebras over the complex field, and the operation J (problem 3.10) on the real tangent spaces induces a bigradation of the forms. As in problem 3.10, we have the various tangent spaces $\mathcal{M}_m$, $\mathcal{T}_m$, and $\mathcal{H}_m$. Since $\mathcal{T}_m = \mathcal{M}_m + i\mathcal{M}_m$, J may be extended to be a complex linear endomorphism of $\mathcal{T}_m$, still satisfying $J^2 = -\mathrm{I}$. The dual of J extends to a derivation J^* of the complex Grassmann algebra $G(\mathcal{T}_m{}^*)$.

$\omega \in G^{p+q}(\mathcal{T}_m{}^*)$ is called an element of *type* (p, q) if $J^*\omega = (p-q) i\omega$. By allowing m to move over M, we define *complex differential forms of type* (p, q). We denote the set of these objects by $\mathcal{H}^{p,q}(M)$. Elements of $\mathcal{H}^{p,0}$ are called *holomorphic* p-forms.

Problem 25. For f a complex-valued C^∞ function on M, define df and show that it is a complex differential 1-form.

Problem 26. For a complex coordinate system $(z_1, \ldots, z_d)$, show that the dz_i, $d\bar{z}_i$ at m, with 1, generate $G(\mathcal{T}_m{}^*)$. Express a form of type (p, q) in terms of these generators, and show that

$$G^r(\mathcal{T}_m{}^*) = \sum_{p+q=r} G^{p,q}(\mathcal{T}_m{}^*).$$

Find dim $G^{p,q}(\mathcal{T}_m{}^*)$.

Problem 27. Show that $\omega \in \mathcal{H}^{p,0}$ if and only if $\omega(X_1, \ldots, X_p) = 0$ whenever $X_1 \in \overline{\mathcal{H}}$.

Problem 28. Show that $d = d' + d''$, where $d' : \mathcal{H}^{p,q} \to \mathcal{H}^{p+1,q}$ and $d'' : \mathcal{H}^{p,q} \to \mathcal{H}^{p,q+1}$. Also verify that $(d')^2 = 0 = (d'')^2$ and $d'd'' = -d''d'$.

Problem 29. Show that the algebras $\mathcal{H}^{p,q}$ may be defined for a manifold with an almost complex structure.

Problem 30. For any almost complex structure,

$$d|_{\mathcal{H}^{0,1}} = d^{2,0} + d^{1,1} + d^{0,2},$$

where $d^{p,q} : \mathcal{H}^{0,1} \to \mathcal{H}^{p,q}$. Show that the condition for the almost complex structure to be a complex structure (see problem 3.11 *et al.*) is equivalent to $d^{2,0} = 0$, and hence to the assertion of problem 28 above.

CHAPTER 5

Connexions

The definition of a connexion on a principal bundle is given and the horizontal lift of a curve and parallel translation are established. The curvature form is defined and the structural equation is proved. Existence of connexions, connexions on associated bundles, and structural equations for horizontal forms are discussed, and holonomy groups are introduced via a sequence of problems [*49, 50, 66, 83*].

5.1 Definitions and First Properties

The appropriate vehicle for certain differential geometric structure on a manifold appears to be a connexion, which in a special case, treated in the next chapter, is equivalent to the formulations in terms of the more usual notion of parallel translation. Connexions are associated with principal bundles over a manifold, and for the most part we shall be considering only connexions on the bundle of bases. However, many fundamental properties of connexions hold on a general principal bundle, and these we develop in this chapter.

Let P be the bundle space of a principal bundle over M with structural group G and projection map $\pi : P \to M$.

Let V be the $\dim(G)$-dimensional distribution on P of *vertical vectors*, that is, for $p \in P$,

$$V_p = \{t \in P_p \mid d\pi \, t = 0\}.$$

Recall that there is a homomorphism $\lambda : \mathfrak{g} \to \bar{\mathfrak{g}}$ defined by the right action of G on P (3.1): if $X \in \mathfrak{g}$, $p \in P$, then $(\lambda X)(p) = dp(X(e))$, where p is here regarded as the injection of G into P given by $p(g) = pg$. Then the elements of $\bar{\mathfrak{g}}$ are vertical, for if $\bar{X} = \lambda X$, then

$$d\pi(\bar{X}(p)) = d\pi \, dp(X(e)) = 0$$

74

since $\pi \circ p$ is the constant map $G \to \pi(p)$. Further, if $t \in V_p$, then there exists an $\bar{X} \in \bar{\mathfrak{g}}$ such that $\bar{X}(p) = t$, since p maps G onto π^{-1} $(\pi(p))$. Hence, the elements of $\bar{\mathfrak{g}}$ span the vertical space at every point, which shows that V is indeed a distribution with dimension equal to $\dim(\mathfrak{g}) = \dim(G)$, and further, $V \in C^\infty$. Indeed, if U is a distinguished neighborhood of $\pi(p)$, F_U the associated map into G, then the map $p: G \to \pi^{-1}(m)$ may be factored as follows: $p = (\pi \times F_U)^{-1}(m, L_{F_U(p)})$, from which follow the desired properties. (See remark near the beginning of 3.2).

A *connexion* on the principal bundle (P, G, M) is a d-dimensional distribution H on P such that

(i) $H \in C^\infty$,

(ii) for every $p \in P$, $H_p + V_p = P_p$, that is, H_p is a linear complement to V_p,

(iii) for every $p \in P$, $g \in G$, $dR_g H_p = H_{pg}$.

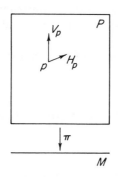

FIG. 16.

If $t \in P_p$, we write $t = Vt + Ht$, where $Vt \in V_p$, $Ht \in H_p$. If X is a vector field, then we write also $(VX)(p) = V(X(p))$, $(HX)(p) = H(X(p))$. Elements of H_p are called *horizontal*.

Problem 1.   Prove that $X \in C^\infty$ if and only if VX and $HX \in C^\infty$.

Since $d\pi \vert_{H_p}$ is one-to-one, $d\pi: H_p \approx M_{\pi(p)}$ by dimensionality. Hence, to every vector field X on M there corresponds a unique vector field $\tilde{X}$ on P, called the *(horizontal) lift of X*, with the properties that for every $p \in P$, $\tilde{X}(p) \in H_p$ and $d\pi\, \tilde{X}(p) = X(\pi(p))$. From problem

1 it follows that if $X \in C^\infty$, then $\tilde{X} \in C^\infty$. We also have

(a) $$dR_g \tilde{X} = \tilde{X}, \qquad g \in G,$$

(b) $$\tilde{X} + \tilde{Y} = \widetilde{(X + Y)}$$

(c) $$H[\tilde{X}, \tilde{Y}] = \widetilde{[X, Y]}.$$

All these properties are trivial to verify.

We now give the dual formulation of a connexion.

The 1-*form of a connexion* H is the Lie algebra valued 1-form ϕ on P defined by:

if $p \in P, t \in P_p$, then $\phi(t) =$ that $X \in \mathfrak{g}$ such that $\tilde{X}(p) = Vt$.

A p-form ω on P is called *vertical (horizontal)* if it vanishes when one or more of its entries is horizontal (vertical). If ω has values in $\mathfrak{g}$, then it is called *equivariant* if, for every $g \in G$, $\omega \circ dR_g = \text{Ad } g^{-1} \circ \omega$.

Notice that the definition of a vertical form depends on the existence of a connexion, that is, a notion of horizontal tangents, whereas the horizontality of a form is a notion that is independent of a connexion.

Lemma 1. The 1-form ϕ of a connexion H has the following properties:

(i) ϕ is vertical,

(ii) If $X \in \mathfrak{g}$, $\tilde{X} \in \bar{\mathfrak{g}}$, then $\phi(\tilde{X}(p)) = X$, all $p \in P$,

(iii) ϕ is equivariant,

(iv) $\phi \in C^\infty$.

Proof. (i) and (ii) are trivial. (iii) follows immediately from 3.1(b). Let X be a C^∞ vector field on P, so there exist functions f_i such that

$$VX = \sum_i f_i \tilde{X}_i,$$

where $\tilde{X}_i \in \bar{\mathfrak{g}}$. Then the $f_i \in C^\infty$ [*25*, prop. 1, p. 88], so $\phi(X) = \Sigma_i f_i X_i \in C^\infty$, proving (iv).

The properties (ii)-(iv) characterize 1-forms of connexions in the following sense.

Theorem 1. If ϕ is a 1-form on P with values in $\mathfrak{g}$ satisfying properties (ii)-(iv), then there exists a unique connexion H on P such that

ϕ is its 1-form. Hence, there is a one-to-one correspondence between such 1-forms on P and connexions on P.

Proof. Let $H_p = \{t \in P_p \mid \phi(t) = 0\}$. Then for $t \in P_p$, let $Vt = \overline{\phi(t)}(p)$, so $t - Vt \in H_p$, which proves property (ii) in the definition of a connexion. Further, for any $g \in G$, $\phi(dR_g t) = \operatorname{Ad} g^{-1}\phi(t) = 0$ if $t \in H_p$, so $dR_g t \in H_{pg}$. Hence, $dR_g H_p \subset H_{pg}$, which proves (iii). Further, if X is a C^∞ vector field on P, then $VX = \overline{\phi(X)} \in C^\infty$, so $HX \in C^\infty$. From this it follows trivially that $H \in C^\infty$, which establishes that H is a connexion.

Because of this theorem we shall sometimes speak of ϕ as being the connexion.

Problem 2. Let ϕ and ψ be two connexion forms on P and f a C^∞ function on M. Show that:

(a) $(f \circ \pi)\phi + (1 - f \circ \pi)\psi$ is a connexion form on P.

(b) If $t \in P_p$ and $t = t_1 + t_2$, $t = {}'t_1 + {}'t_2$ are the horizontal and vertical decompositions of t with respect to the connexions of ϕ and ψ, find the decomposition with respect to the connexion of $(f \circ \pi)\phi + (1 - f \circ \pi)\psi$. [*Hint:* $\phi({}'t_1) = -\psi(t_1)$.] In particular, when we take $f = $ constant, this shows that the connexions of P form an affine space.

5.2 Parallel Translation

If γ is a broken C^∞ curve in M, then a (*horizontal*) *lift* of γ is a broken C^∞ curve $\tilde{\gamma}$ in P such that (i) $\tilde{\gamma}$ is horizontal, that is, $\tilde{\gamma}_*$ is horizontal, and (ii) $\pi \circ \tilde{\gamma} = \gamma$. By *broken* we mean γ is continuous and piecewise C^∞.

Theorem 2. Let γ be a broken C^∞ curve in M, $\gamma : [0, 1] \to M$. Let $p \in \pi^{-1}(\gamma(0))$. Then there exists a unique lift $\tilde{\gamma}$ of γ such that $\tilde{\gamma}(0) = p$.

Proof. We may assume that γ is C^∞ since we may chain the lifts of pieces together, in fact, in only one way.

Extend γ to a C^∞ map of $(-\epsilon, 1 + \epsilon) = I$ into M. Then by problem 3.6, $N = \{(r, q) \in I \times P \mid \gamma(r) = \pi(q)\}$ is the bundle space of a principal bundle (N, G, π', I). The map $\theta : N \to P$ by $\theta(r, q) = q$ may be used to get a connexion on N; the 1-form of this connexion is $\check{\phi} = \theta^*\phi$ (see problem 3 below).

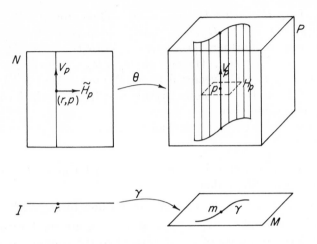

FIG. 17.

Let X be the unique horizontal lift of D on I to N, and let $\tilde{u}$ be the integral curve of X starting at $(0, p)$. The domain of $\tilde{u}$ is all of I since it could be extended about a neighborhood of the upper limit otherwise.

We define $\tilde{\gamma} = \theta \circ \tilde{u}$. It is obvious that $\tilde{\gamma}$ is a lift of γ, and since $\phi(\tilde{\gamma}_*) = \phi(d\theta \circ \tilde{u}_*) = \check{\phi}(X) = 0$, $\tilde{\gamma}$ is horizontal.

That $\tilde{\gamma}$ is unique follows easily from the fact that any lift can be factored through N via θ and a lift of u, but the latter is obviously unique. QED

Corollary 1. If H is a connexion on (P, G, M), γ a curve in M as in the theorem, then there is defined a diffeomorphism T_γ of $\pi^{-1}(\gamma(0))$ onto $\pi^{-1}(\gamma(1))$, called *parallel translation from $\gamma(0)$ to $\gamma(1)$ along γ*. T_γ is independent of the parametrization of γ and satisfies $T_\gamma \circ R_g = R_g \circ T_\gamma$, all $g \in G$. Further, if γ and σ are two such curves with $\sigma(0) = \gamma(1)$, then $T_{\gamma\sigma} = T_\sigma \circ T_\gamma$.

Proof. Let $p \in \pi^{-1}(\gamma(0))$, $\tilde{\gamma}$ be the lift of γ with $\tilde{\gamma}(0) = p$, and define $T_\gamma(p) = \tilde{\gamma}(1)$. We use the right invariance property, which is trivial, to prove $T_\gamma \in C^\infty$ and T_γ is a diffeomorphism: if p_0 is any fixed element of $\pi^{-1}(\gamma(0))$, $p_1 = T_\gamma(p_0)$, then $T_\gamma(p_0 g) = T_\gamma(p_0) g$, so

$$T_\gamma = p_1 \circ p_0^{-1} : \pi^{-1}(\gamma(0)) \to G \to \pi^{-1}(\gamma(1)).$$

The other properties are immediate.

We remark that a concept of parallel translation is equivalent to a connexion; namely, if we have parallel translation in P given and satisfying right invariance and certain smoothness conditions—specifically, tangent curves give coincident infinitesimal transformations—then the connexion may be recovered by differentiating as follows: let γ be an appropriate curve in M, $p \in \pi^{-1}(\gamma(0))$, and let $t \in P_p$ such that $d\pi\, t = \gamma_*(0)$. Then define $Ht = T_\gamma(p)_*(0)$, where we view $T_\gamma(p)$ as the curve $t \to$ parallel translate of p from $\gamma(0)$ to $\gamma(t)$ along γ.

The following result provides an interesting interpretation of the connexion form.

Corollary 2. Let γ, $\tilde{\gamma}$ be as in theorem 2, and let τ be any other lift (not necessarily horizontal). Define $\alpha : I \to G$ by: $\alpha(r)$ is the unique element of G such that $\tilde{\gamma}(r) = \tau(r)\alpha(r)$.

Then $dR_{(\alpha(r)^{-1})}\alpha_*(r) = -\phi(\tau_*(r))$. (We are identifying G_e with $\mathfrak{g}$ by left translation.)

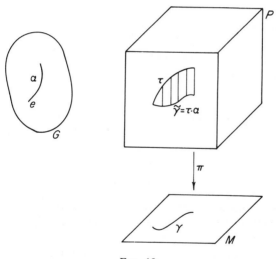

Fig. 18.

Proof. Let $I_r : G \to P$ by $I_r(g) = \tau(r)\,g$. It is trivial to verify that $I_r \circ L_g = \tau(r)\,g : G \to P$. By theorem 1.2, where the map $P \times G \to P$ is right action, we have

$$\tilde{\gamma}_*(r) = dI_r\alpha_*(r) + dR_{\alpha(r)}\,\tau_*(r). \tag{1}$$

Now

$$dI_r \alpha_*(r) = dI_r dL_{\alpha(r)} \, dL_{\alpha(r)^{-1}}(\alpha_*(r))$$

$$= d(\tau(r) \, \alpha(r))(dL_{\alpha(r)^{-1}}(\alpha_*(r)))$$

$$= \lambda(dL_{\alpha(r)^{-1}}(\alpha_*(r)))(\tau(r) \, \alpha(r)) \qquad \text{(3.1 and definition of } \lambda X\text{)}.$$

Therefore, $\phi(dI_r \alpha_*(r)) = dL_{\alpha(r)^{-1}}(\alpha_*(r))$, recalling our identification of G_e and $\mathfrak{g}$.

On the other hand, $\phi(dR_{\alpha(r)} \tau_*(r)) = \text{Ad } \alpha(r)^{-1} \, \phi(\tau_*(r))$, by the equivariance of ϕ. Hence, since $\tilde{\gamma}$ is horizontal, applying ϕ to (1) gives

$$dL_{\alpha(r)^{-1}}(\alpha_*(r)) = -\text{Ad } \alpha(r)^{-1} \phi(\tau_*(r))$$

$$= - dL_{\alpha(r)^{-1}} \circ dR_{\alpha(r)}(\phi(\tau_*(r))),$$

so

$$dR_{\alpha(r)^{-1}}(\alpha_*(r)) = -\phi(\tau_*(r)). \qquad \text{QED}$$

Problem 3. Let $(f_B, f_G, f_M) : (B, G, M) \to (B', G', M')$ be a bundle map with $df_G : \mathfrak{g} \to \mathfrak{g}'$ an isomorphism onto. Show that any connexion on B' induces in a natural way a connexion on B. In particular, if $f: M \to M'$ and (B, G', M) is the bundle induced by f over M, then this will be the case.

5.3 Curvature Form and the Structural Equation

Define, for a form ω on P, the form $D\omega$ by

$$D\omega = d\omega \circ H,$$

where H is a connexion. More precisely, if ω is a p-form, then for $t_1, \ldots, t_{p+1} \in P_e$, $D\omega(t_1, \ldots, t_{p+1}) = d\omega(Ht_1, \ldots, Ht_{p+1})$. Note that $D\omega$ is always horizontal.

The *curvature form* Φ of a connexion H with 1-form ϕ is the horizontal $\mathfrak{g}$-valued 2-form $D\phi$. It is easy to verify that Φ is equivariant.

We need the following lemma for the proof of Cartan's structural equation.

Lemma 2. If $X \in \mathfrak{g}$, V a horizontal vector field on P, then $[\bar{X}, V]$ is horizontal.

Proof.   We cannot apply theorem 3.1 directly, since the horizontal vector fields do not form a finite-dimensional vector space. However, we may derive the result indirectly as follows. Let V_i be a right invariant horizontal vector field, that is, a horizontal lift of a vector field on M. Then, taking $\mathscr{L} = \{V_i\}$ in theorem 3.1, we have $[\bar{X}, V_i] = 0$. Now locally we may write $V = \Sigma_i f_i V_i$, where the V_i are horizontal and right invariant and the f_i are C^∞ functions on P. Then $[\bar{X}, V] = \Sigma_i(\bar{X}f_i) V_i$, by problem 1.14, which is certainly horizontal.

Problem 4.   Applying the formula $L_{\bar{X}} = i(\bar{X})\, d + di(\bar{X})$ to the connexion form ϕ and evaluating on the horizontal vector field V, show that $\phi([\bar{X}, V]) = 0$, thus giving an alternate proof of the lemma. [*Hint:* Notice that the one-parameter group of transformations associated with $\bar{X}$ is right action by e^{tX}. Hence $L_{\bar{X}}(\phi)$ is vertical.]

Theorem 3   (*structural equation*).   If ϕ is the 1-form of a connexion on P, Φ its curvature form, then

$$d\phi = -\tfrac{1}{2}[\phi, \phi] + \Phi.$$

   Notice that if G is a matrix group and $\mathfrak{g}$ is identified with a space of linear transformations, then $-\tfrac{1}{2}[\phi, \phi] = -\phi^2$.

Proof.   We show that the above 2-forms applied to vector fields X, Y on P agree. Since forms are linear, we need only consider the cases for which the X, Y either belong to $\bar{\mathfrak{g}}$ or are horizontal, in all combinations.

   (i) X, $Y \in \bar{\mathfrak{g}}$, so there are elements X', $Y' \in \mathfrak{g}$ such that $\bar{X}' = X$, $\bar{Y}' = Y$, and hence $\phi(X) = X'$, $\phi(Y) = Y'$. Now by theorem 4.2

$$d\phi(X, Y) = X\phi(Y) - Y\phi(X) - \phi([X, Y])$$
$$= X(Y') - Y(X') - [X', Y']$$
$$= -\tfrac{1}{2}[\phi, \phi](X, Y)$$
$$= -\tfrac{1}{2}[\phi, \phi](X, Y) + \Phi(X, Y), \qquad (4.9)$$

as desired, since Φ is horizontal. Note that $X(Y') = X$(constant $\mathfrak{g}$-valued function on P) $= 0$, and $Y(X') = 0$ similarly.
   (ii) $X \in \bar{\mathfrak{g}}$, Y horizontal. Let $X' \in \mathfrak{g}$ be as in (i):

$$d\phi(X, Y) = X\phi(Y) - Y\phi(X) - \phi([X, Y]) = 0,$$

since $\phi(Y) = 0$, $\phi(X) =$ constant, and $[X, Y]$ is horizontal by the lemma. On the other hand, Φ is horizontal, so $\Phi(X, Y) = 0$ as X is vertical, and $[\phi, \phi](X, Y) = 0$ since Y is horizontal.

(iii) X, Y both horizontal. $\phi(X) = 0$, $\phi(Y) = 0$, and

$$\Phi(X, Y) = d\phi(HX, HY) = d\phi(X, Y),$$

which is the structural equation in this case. QED

Remark. The restriction of the structural equation to vertical vectors is essentially the equation of Maurer-Cartan. Another interpretation is that it says that $d\phi$ has only a horizontal and a vertical part with no *mixed part*.

Theorem 4 (*the Bianchi identity*). If Φ is the curvature form of a connexion on a principal bundle P, then

$$D\Phi = 0.$$

Proof. From the structural equation,

$$D\Phi = D\, d\phi - \tfrac{1}{2} D[\phi, \phi].$$

Now $Dd\phi(X_1, X_2, X_3) = dd\phi(HX_1, HX_2, HX_3) = 0$, since $d^2 = 0$. Also, $D[\phi, \phi] = 0$, since $[\phi, \phi]$ is a vertical 2-form, and so vanishes when one of its entries is horizontal. Hence, $D\Phi = 0$.

Theorem 5. Let H be a connexion on P, Φ its curvature form. Then $\Phi = 0$ if and only if H is an involutive distribution, which in view of theorem 1.6, means that P admits local horizontal cross sections. In particular, if M is simply connected, then by a standard monodromy argument P must be the trivial bundle. A connexion with $\Phi = 0$ is called *flat*.

Proof. If X, Y are horizontal vector fields on P, then

$$\Phi(X, Y) = d\phi(X, Y) = X\phi(Y) - Y\phi(X) - \phi([X, Y])$$
$$= -\phi([X, Y]);$$

so H is involutive if and only if $[X, Y]$ is horizontal if and only if $\Phi(X, Y) = -\phi([X, Y]) = 0$, which implies $\Phi = 0$, as asserted.

Problem 5. Let H be a closed subgroup of a Lie group G and consider the principal bundle $(G, H, G/H)$. Let ϕ be a connexion and show that $\phi \circ dR_h = dR_h \circ \phi$. Now assume ϕ is invariant under dL_g for every $g \in G$. Show that

(a) ϕ defines a projection $\bar{\phi} : \mathfrak{g} \to \mathfrak{h}$ of the corresponding Lie algebras,

(b) if $\mathfrak{m} = \ker \bar{\phi}$, then $[\mathfrak{m}, \mathfrak{h}] \subset \mathfrak{m}$,

(c) conversely, if $\mathfrak{g} = \mathfrak{m} + \mathfrak{h}$ (direct sum) with $[\mathfrak{m}, \mathfrak{h}] \subset \mathfrak{m}$, then the projection of $\mathfrak{g}$ onto $\mathfrak{h}$ gives rise to an invariant connexion in the above sense. Hence, there is a one-to-one correspondence between invariant connexions on $(G, H, G/H)$ and *reductive* complements $\mathfrak{m}$ of $\mathfrak{h}$. An H admitting such an invariant complement $\mathfrak{m}$ is called *reductive in G* (the name arises from the fact that the adjoint representation of G restricted to H is reducible to the adjoint representation of H plus the representation of H on $\mathfrak{m}$ via Ad_G, at least if H is connected).

(d) For the connexion ϕ, show that the curvature form may be considered as defined on $\mathfrak{g} \times \mathfrak{g}$, and derive a formula for it.

5.4 Existence of Connexions and Connexions in Associated Bundles

Existence of Connexions. C^∞ connexions exist in abundance. In Chapter 7 we shall establish the existence of Riemannian connexions on $B(M)$. Here we show that any principal bundle (P, G, M), with M paracompact, has a connexion.

Let $\{U_i\}$ be a covering of M such that $\pi^{-1}(U_i)$ is trivial. Let $\{f_i\}$ be a C^∞ partition of unity subordinate to the covering $\{U_i\}$. Let ϕ_i be a flat connexion on $\pi^{-1}(U_i)$ and define $\phi = \Sigma (f_i \circ \pi) \phi_i$. ϕ is a not necessarily flat connexion form on P.

Problem 6. Verify that ϕ is a connexion form.

Remark. If (P, G, π, M) is a complex analytic principal bundle over a complex manifold M, it of course admits C^∞ connexions, but in general it will not admit a complex analytic connexion. A necessary condition in a special case is given in [3].

However, real-analytic manifolds admit analytic connexions, although the proof is much more difficult. See the remark following theorem 7.2.

Associated Bundles. Let (P, G, M) be a principal bundle with a connexion H, and let (B, G, F, M) be an associated bundle with fibre

F (see 3.3). Then in some sense H induces a "connexion" on B. To be precise, there is a distribution H' on B which at each point complements the vertical tangent space. Further, there is a notion of parallel translation of the fibres of B, which derives as before from the horizontal lifts of curves.

Parallel Translation. Let γ be a broken C^∞ curve in M, $b \in \pi'^{-1}(\gamma(0))$. We define a lifting $\bar{\gamma}$ of γ into B which will turn out to be horizontal in the sense below. Let $f \in F$ and $p \in P$ be such that $\pi(p) = \gamma(0)$ and $pf = b$, where p is here the map defined in 3.3. By theorem 2 there is a horizontal lifting $\tilde{\gamma}$ of γ into P with $\gamma(0) = p$. Now define $\bar{\gamma}$ by $\bar{\gamma}(t) = \tilde{\gamma}(t) f$. We then define *parallel translation* T_γ along γ from $\pi'^{-1}(\gamma(0))$ to $\pi'^{-1}(\gamma(1))$ as in P. Hence we have $T_\gamma = T_\gamma(p) \circ p^{-1}$, so parallel translation is a diffeomorphism.

The Distribution H'. Take $b \in B$, $p \in P$ such that $\pi'(b) = \pi(p)$. We may view P_p as a subspace of $(P \times F)_{(p,f)}$, where $f \in F$ is such that $pf = b$. Let $\lambda: P \times F \to B$ be the natural map (3.3), and define $H'_b = d\lambda(H_p)$. This definition is independent of p in view of the right invariance of H, while it is clear that the lift defined above is horizontal with respect to H', if the definition of the map

$$p : F \to \pi'^{-1}(\pi(p))$$

is recalled.

Problem 7. Let ϕ, ψ be connexion forms, H, K their connexions, H', K' the corresponding distributions on B. Show that if s, $t \in B_b$, $H's = s$, $K't = t$, and $d\pi'(s) = d\pi'(t)$, then $rs + (1 - r)\,t$ is in the distribution on B of the connexion belonging to $r\phi + (1 - r)\,\psi$.

Problem 8. Determine all connexions on $T(R)$, the tangent bundle to R.

Problem 9. Show that there exist horizontal distributions on associated bundles which are not connexions. [*Hint*: Take $T(R) \approx R^2$ and define a distribution with slope e^y.]

5.5 Structural Equations for Horizontal Forms

We first prove a basic lemma.
Let G be a Lie group of diffeomorphisms of M, $G \times M \to M$,

and let ϕ be a representation of G as nonsingular linear transformations on a vector space V. Then there is an associated representation $\tilde{\phi}$ of $\mathfrak{g}$ as linear transformations on V, which may be defined as follows. If $v \in V$, $X \in \mathfrak{g}$, then set $\tilde{\phi}(X) v = X(e) v$. This makes sense since v is a vector-valued function on G, namely, $v(g) = \phi(g) v$; and so is mapped into a vector by the tangent $X(e)$ (see 1.4, 2.2, and 2.6).

Lemma 3. Let $X \in \mathfrak{g}$ and ω be a V-valued p-form, and let $Y_1, ..., Y_p$ be invariant vector fields on M.

(i) If ω satisfies $\omega \circ dg = \phi(g) \omega$ for all $g \in G$, then

$$(\tilde{\phi}X) \omega(Y_1, ..., Y_p) = (\lambda X) \omega(Y_1, ..., Y_p),$$

where λX is the vector field on M defined in 3.1.

(ii) If ω satisfies $\omega \circ dg = \phi(g^{-1}) \omega$ for all $g \in G$, then

$$-(\tilde{\phi}X) \omega(Y_1, ..., Y_p) = (\lambda X) \omega(Y_1, ..., Y_p).$$

Proof. We shall prove (ii). The proof of (i) is similar. If $f \in M$, then $\omega(Y_1, ..., Y_p) \circ f$ is a V-valued function on G, and in fact

$$\omega(Y_1, ..., Y_p) \circ f(g) = \omega(Y_1(gf), ..., Y_p(gf)) = \omega(dgY_1(f), ..., dgY_p(f))$$
$$= \phi(g^{-1}) \omega(Y_1(f), ..., Y_p(f))$$
$$= \omega(Y_1(f), ..., Y_p(f)) \psi(g),$$

where $\psi(g) = g^{-1}$, since the Y_i are invariant and ω is equivariant with respect to the representation ϕ. Therefore we have

$$(\lambda X) \omega(Y_1, ..., Y_p)(f) = (\lambda X)(f)(\omega(Y_1, ..., Y_p))$$
$$= df\, X(e)(\omega(Y_1, ..., Y_p)) \qquad (3.1)$$
$$= X(e)(\omega(Y_1, ..., Y_p) \circ f)$$
$$= X(e)(\omega(Y_1(f), ..., Y_p(f)) \circ \psi)$$
$$= d\psi\, X(e)(\omega(Y_1(f), ..., Y_p(f)))$$
$$= -X(e)(\omega(Y_1(f), ..., Y_p(f))) \qquad \text{(Lemma 2.1)}$$
$$= -\tilde{\phi}(X) \omega(Y_1, ..., Y_p)(f),$$

the last step following from the definition of $\tilde{\phi}$. This gives the desired result.

We apply this lemma to the case of a principal bundle (P, G, M), that is, G acts to the right on P.

Theorem 6.　Let ϕ be the 1-form of a connexion on P and let ω be a $\mathfrak{g}$-valued, horizontal, equivariant p-form on P. Then we have the structural equation for ω:

$$d\omega = -[\phi, \omega] + D\omega.$$

Proof.　We prove this by applying both sides to $p + 1$ vector fields $Y_1, ..., Y_{p+1}$ chosen from a set which locally spans the tangent space to P. For this set we choose vector fields $\{\lambda X\}$, $X \in \mathfrak{g}$, to span the vertical tangent space. The remaining vector fields we choose in various ways. We consider several cases.

(i) *No Y_i is vertical.* We may then assume the Y_i are all horizontal, and so $[\phi, \omega](Y_1, ..., Y_{p+1}) = 0$ since ϕ is vertical. Also, $HY_i = Y_i$, so $D\omega(Y_i, ..., Y_{p+1}) = d\omega(Y_1, ..., Y_{p+1})$, which proves the result in this case.

(ii) *One Y_i is vertical.*　Assume $Y_{p+1} = \lambda X$. We may choose $Y_1, ..., Y_p$ so that they are right invariant and so that $[Y_i, \lambda X] = 0$. To do this in the neighborhood of a point $f \in P$, we choose a coordinate system at f which derives from the local product structure of P. Then the partial derivatives with respect to the variables coming from M suffice, for they are clearly right invariant and they bracket correctly with λX, since λX depends only on the other coordinates. Now $H(\lambda X) = 0$ implies that $D\omega(Y_1, ..., Y_{p+1}) = 0$. We also have from theorem 4.2

$$d\omega(Y_1, ..., Y_{p+1}) = \sum_i (-1)^{i-1} Y_i \omega(Y_1, ..., Y_{i-1}, Y_{i+1}, ..., Y_{p+1})$$

$$+ \sum_{i<j} (-1)^{i+j} \omega([Y_i, Y_j], Y_1, ..., Y_{i-1}, Y_{i+1}, ..., Y_{j-1}, Y_{j+1}, ..., Y_{p+1})$$

$$= (-1)^p Y_{p+1}\omega(Y_1, ..., Y_p) \quad \text{since } \omega \text{ is horizontal}$$

$$= (-1)^p(\lambda X)\,\omega(Y_1, ..., Y_p)$$

$$= (-1)^{p+1} \operatorname{ad} X\,\omega(Y_1, ..., Y_p)$$

by the lemma with $\phi = \operatorname{Ad}$ and 2.6, since ω is equivariant,

$$= (-1)^{p+1}[X, \omega(Y_1, ..., Y_p)]$$

$$= (-1)^{p+1}[\phi(\lambda X), \omega(Y_1, ..., Y_p)]$$

$$= -[\phi, \omega](Y_1, ..., Y_p, Y_{p+1}),$$

using 4.9 and fact that ω is horizontal. This is what we wanted to prove.

(iii) *Two or more* Y_i *vertical.* From the fact that ω is horizontal it is clear that everything vanishes. QED

Corollary. If Φ is the curvature form associated with ϕ, then we have

$$d\Phi = -[\phi, \Phi].$$

Proof. Φ is horizontal and equivariant, and so the result follows from the theorem and Bianchi's identity.

We shall have several more occasions for employing the above lemma in the ensuing chapters.

5.6 Holonomy [2, 5, 51, 66, 77, 82]

We develop the material in this section as a series of problems.

Let (P, G, π, M) be a principal bundle with connexion ϕ. Let $p \in P$ and define $K_p \subset G$ by $K_p = \{g \in G | pg$ is a parallel translate of $p\}$.

Problem 10. K_p is a subgroup of G, the *holonomy subgroup* of ϕ at p.

Problem 11. If $p' \in P$ is a parallel translate of p, then $K_p = K_{p'}$.

Problem 12. If $p' = pg$, $g \in G$, $K_{p'} = K_{pg} = g^{-1}K_pg$. (Hence, the *holonomy group* of ϕ is defined up to isomorphism.)

Problem 13. Let $K_{p0} = \{g \in K_p \mid pg$ is a parallel translate of p along a null homotopic curve}. Show that K_{p0} is an arcwise connected normal subgroup of K_p and hence an arcwise connected subgroup of G, called the *restricted holonomy group* of ϕ at p. It then follows from a theorem of Yamabe [95] that K_{p0} is a Lie subgroup of G.

Problem 14. There is a natural homomorphism of the fundamental group of M based at $\pi(p)$ onto the quotient group K_p/K_{p0}.

Problem 15. If M is simply connected, then $K_{p0} = K_p$. For example, if $M = R$, then $K_p = \{1\}$ since every connexion is integrable.

It is known that K_{p0} is the component of the identity of K_p [66] so that K_p is a Lie group. Let $P_p' = \{p' \in P \mid p'$ is a parallel translate of $p\}$.

Problem 16. $(P_p', K_p, M, \pi \mid_{p}')$ is a principal bundle, and the inclusion $i_p : P_p' \to P$ gives a reduction of G to K_p. Furthermore, $i_p*(\phi)$ is a connexion form on P_p'.

Problem 17. If G can be reduced to a subgroup K via a bundle map $i : P' \to P$ such that $i*(\phi)$ is a connexion on (P', K, M), then for any $p \in P'$, $K_p \subset K$.

Problem 18. If Φ is the curvature form of ϕ, then for every $p \in P$, $\Phi(P_p, P_p) \subset \mathfrak{k}_p$, the Lie algebra of K_p. (Ambrose-Singer [2] prove even more, namely, if $V = \{\Phi(P_p, P_p) | p$ is a parallel translate of a single $p_0 \in P\}$, then the Lie algebra generated by V is $\mathfrak{k}_{p_0}$. A later result [70] shows that V spans $\mathfrak{k}_{p_0}$.)

Problem 19. Any discrete subgroup of the positive real numbers can be realized as the holonomy group of a connexion on the bundle of bases $B(S^1)$ of the circle, and no other subgroup can arise as a holonomy group.

Problem 20. If ϕ is a connexion on (S^{2d+1}, S^1, CP^d), then the holonomy group at any point is S^1.

Problem 21. If ϕ is a connexion on (S^{4d+3}, S^3, QP^d), then the holonomy group is either S^3 or S^1.

In the above problems it will be found necessary to show that any two broken C^∞ loops which are homotopic are homotopic via a homotopy which is broken C^∞ at every stage, and that any homotopy class of loops has a broken C^∞ member.

CHAPTER 6

Affine Connexions

The additional structure available for a connexion on the bundle of bases is defined, including the torsion form, basic vector fields, the torsion and curvature transformations, and geodesics. The additional structural equation is proved and difference forms are considered, particularly in their relation to torsion and the configuration of geodesics. The exponential map, completeness, and normal coordinates are defined. The chapter concludes with a treatment of covariant differentiation and the classical definitions of the above [*33, 49, 50, 66, 83*].

6.1 Definitions

Let M be a manifold, $B(M)$ its bundle of bases. Then a connexion on $B(M)$ is called an *affine connexion*. Since any connexion on a subbundle of $B(M)$ can be extended to an affine connexion by the right action of the group $GL(d, R)$, such a connexion is also called an affine connexion.

From the notion of parallel translation in the bundle $B(M)$ given by an affine connexion we obtain a notion of *parallel translation in the tangent bundle*, or *parallel translation of tangents along curves*. Since the tangent bundle $T(M)$ is an associated bundle of $B(M)$, we can derive this property from 5.4, but it is simple enough to give an explicit definition in this case.

If γ is a curve in M and $t \in M_m$, $m = \gamma(0)$, then we obtain the parallel translate of t along γ to $n = \gamma(u)$ as follows: choose any $b \in B(M)$ with $\pi(b) = m$, and let $\bar{\gamma}$ be the unique horizontal lifting of γ to b. Then if $\bar{\gamma}(s) = (\gamma(s), e_1(s), ..., e_d(s))$ and $t = \Sigma_i a_i e_i(0)$, we define the parallel translate of t to be $\Sigma a_i e_i(u)$. It is easy to check, using the invariance of the connexion under right action, that this transformation is independent of the choice of b over m.

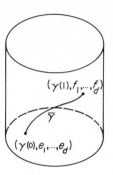

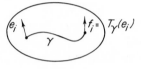

FIG. 19.

6.1.1 The Solder Forms

$B(M)$ always has defined on it certain horizontal 1-forms, which are independent of any connexion on the bundle.

The *solder* 1-*forms* ω_i are defined as follows. Let $t \in B(M)_b$, where $b = (m, e_1, ..., e_d)$. Then

$$d\pi\, t = \sum \omega_i(t)\, e_i.$$

Or these ω_i can be considered as a single R^d-valued 1-form ω, defined by

$$\omega(t) = (\omega_1(t), ..., \omega_d(t)).$$

Lemma 1. The solder form satisfies the following properties:

(i) $\omega \in C^\infty$,

(ii) ω is horizontal,

(iii) ω is *equivariant*, that is, for every $g \in Gl(d, R)$,

$$R_g{}^*(\omega) = \omega \circ dR_g = g^{-1} \circ \omega,$$

where on the right-hand side g is viewed as acting to the left in R^d.

Proof. (ii) is immediate. To prove (iii), let

$$t \in B(M)_b, \, b = (m, e_1, ..., e_d), \, d\pi \, t = \sum a_i e_i \, .$$

Then $dR_g t \in B(M)_{bg}$, $bg = (m, \Sigma_i g_{i1} e_i, ..., \Sigma_i g_{id} e_i)$. Now

$$d\pi \, t = \sum_{i,j,k} (g_{ij}^{-1} a_j)(g_{ki} e_k),$$

so

$$\omega_i(dR_g t) = \sum g_{ij}^{-1} a_j.$$

Hence, $\omega(dR_g t) = g^{-1}\omega(t)$, as asserted.

The proof of (i) is direct. Let y_i, y_{jk} be product coordinates on a coordinate neighborhood of $B(M)$ (3.2), and we have to show that $\omega(D_{y_{jk}})$ and $\omega(D_{y_i})$ are C^∞. By (ii), $\omega(D_{y_{jk}}) = 0$, so we need only consider $\omega(D_{y_i})$. But

$$d\pi \, D_{y_i}(b) = \sum_j y_{ji}^{-1}(b) \, e_j \qquad \text{(problem 3.4)}$$

where $b = (m, e_1, ..., e_d)$. Hence,

$$\omega(D_{y_i}) = (y_{1i}^{-1}, ..., y_{di}^{-1}) \in C^\infty. \qquad \text{QED}$$

6.1.2 Fundamental and Basic Vector Fields

A vector field $\bar{X}$ on $B(M)$ is called *fundamental* if $\bar{X} \in \lambda(\mathfrak{gl}(d, R))$, that is, if there is an $X \in \mathfrak{gl}(d, R)$ such that $\bar{X} = \lambda X$, (3.1). In particular, the fundamental vector field corresponding to $X_{ij} \in \mathfrak{gl}(d, R)$, where X_{ij} is the matrix with 1 in the (i, j)th entry and 0 elsewhere, is denoted by E_{ij}.

The following lemma summarizes the properties of a fundamental vector field.

Lemma 2. If $\bar{X}$ is a fundamental vector field on $B(M)$, then we have

(i) $\bar{X} \in C^\infty$,
(ii) $\bar{X}$ is vertical,
(iii) if $\bar{X} = \lambda X$, then

$$dR_g \bar{X} = \lambda(\text{Ad } g^{-1} X) \qquad [3.1(b)]$$

for $g \in Gl(d, R)$.

Let H be a connexion on $B(M)$, and let $x \in R^d$, $b \in B(M)$. Then there is a unique horizontal tangent $E(x)(b)$ at b such that $\omega(E(x)(b)) = x$, since $d\pi$ is an isomorphism on H_b. The vector field on $B(M)$ whose value at b is $E(x)(b)$ is called a *basic vector field* and denoted

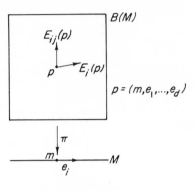

FIG. 20.

by $E(x)$. In particular, if $\delta_i = (\delta_{1i}, ..., \delta_{di}) \in R^d$, then we have basic vector fields $E_i = E(\delta_i)$ which give a basis of the horizontal tangent space at every point of $B(M)$. These $E(x)$ are not the horizontal lifts of any vector fields on the base space M, not being right invariant by (iii) of this lemma.

Lemma 3. Let $E(x)$ be a basic vector field on $B(M)$. Then we have

(i) $E(x) \in C^\infty$,
(ii) $E(x)$ is horizontal (by definition),
(iii) $dR_g E(x) = E(g^{-1}x)$,

for $g \in Gl(d, R)$, where g is viewed as acting on R^d to the left, as in lemma 1.

Parts (i) and (iii) follow from lemma 1 and the (defining) fact that $\omega(E(x)) = x$.

Notice that the vector fields E_i, E_{jk} give a *parallelization* of $B(M)$, that is, for every $b \in B(M)$, $E_i(b)$, $E_{jk}(b)$ are a basis of $B(M)_b$. Further, they are dual to the 1-forms ω_i, ϕ_{jk}, where ϕ_{jk} is the 1-form defined by $\phi_{jk}(t) = (j, k)$th entry of $\phi(t)$, ϕ the 1-form of the connexion. E_{jk} and ω_i do not depend on the connexion but are intrinsic on the bundle of bases. E_i and ϕ_{jk} on the other hand do depend on the connexion, and in fact the connexion H may be given by specifying either the $\{E_i\}$ or the $\{\phi_{jk}\}$. For the ϕ_{jk} we saw this in theorem 5.1.

Moreover, if E_i are d linearly independent nonvanishing vector fields on $B(M)$ satisfying $\omega(E_i) = \delta_i$ then the distribution H defined by setting

$$H_b = \text{span of the } \{E_i(b)\}$$

is clearly a connexion on $B(M)$ with basic vector fields E_i.

6.1.3 Alternate Definition of the Solder Form

Sometimes it is convenient to regard each $b \in B(M)$ as an isomorphism of R^d onto M_m, where $b = (m, e_1, ..., e_d)$, by defining

$$b(x) = \sum x_i e_i .$$

This corresponds to the fact that $T(M)$, the tangent bundle to M, is an associated bundle to $B(M)$ under the left action of $Gl(d, R)$ on R^d, so that we have $b(gx) = bg(x)$ for $g \in G$ (see 3.3).

Lemma 4. Let $b \in B(M)$, $t \in B(M)_b$, then

$$\omega(t) = b^{-1}(d\pi\, t),$$

and this formula may be used to define ω.
Further, if $x \in R^d$, then

$$E(x)(b) = (d\pi\, |_{H_b})^{-1}(bx).$$

The proofs of these statements are immediate, and the advantage of using these formulas as definitions is that they tend to put these concepts and subsequent manipulations of them on a more intrinsic and subscript-free footing. For example, property (iii) of lemma 1 might be proved as follows. The mapping $bg : R^d \xrightarrow{\ g\ } R^d \xrightarrow{\ b\ } M_m$ has an inverse

$$(bg)^{-1} = g^{-1}b^{-1} : M_m \xrightarrow{\ b^{-1}\ } R^d \xrightarrow{\ g^{-1}\ } R^d,$$

and hence,

$$\omega(dR_g t) = (bg)^{-1}(d\pi\, t) = g^{-1}(b^{-1}(d\pi\, t)) = g^{-1}\omega(t).$$

6.1.4 Torsion

The *torsion form* Ω of an affine connexion H on $B(M)$ is the R^d-valued 2-form

$$\Omega = D\omega = d\omega \circ H.$$

(Compare the curvature form.) It is easy to verify that Ω is C^∞, horizontal, and equivariant, that is,

$$\Omega \circ dR_g = g^{-1} \circ \Omega.$$

We now relate torsion and curvature forms to the basic vector fields of the connexion.

Lemma 5. Let x, $y \in R^d$. Then we have

$$[E(x), E(y)] = -\lambda\Phi(E(x), E(y)) - E(\Omega(E(x), E(y)));$$

that is, the curvature and torsion give the vertical and horizontal components of the bracket of two basic vector fields.

Proof. We show that ω and ϕ applied to each side yield the same function, and this will be sufficient since these forms are parallelizing, that is, are dual to a set of parallelizing vector fields. The right-hand side is easy to calculate since λX is vertical and $E(z)$ is horizontal:

$$\phi(\text{right side}) = -\phi(\lambda\Phi(E(x), E(y)))$$
$$= -\Phi(E(x), E(y)),$$
$$\omega(\text{right side}) = -\omega(E(\Omega(E(x), E(y)))$$
$$= -\Omega(E(x), E(y)).$$

To apply ϕ and ω to the left-hand side we use the intrinsic formulas for the exterior derivatives $d\phi$, $d\omega$ (see 4.6).

$$-\phi(\text{left side}) = -\phi([E(x), E(y)])$$
$$= E(x)\,\phi(E(y)) - E(y)\,\phi(E(x)) - \phi([E(x), E(y)])$$
$$= d\phi(E(x), E(y))$$
$$= d\phi(HE(x), HE(y))$$
$$= D\phi(E(x), E(y))$$
$$= \Phi(E(x), E(y)).$$

Similarly,

$$-\omega(\text{left side}) = d\omega(E(x), E(y))$$
$$= \Omega(E(x), E(y)).$$

We have used the facts that $\phi(E(y)) = \phi(E(x)) = 0 =$ constant and $\omega(E(x)) = x =$ constant, $\omega(E(y)) = y =$ constant, so that their derivatives in the directions $E(y)$ and $E(x)$ are 0. QED

Theorem 1. Let ϕ be a connexion form on $B(M)$. Then the curvature and torsion forms of ϕ vanish on $B(M)$ if and only if the following condition on M is satisfied.

Let $m \in M$. There exists a coordinate system $(x_1, \ldots, x_d)$ at m, with domain U, such that the image of the cross section $\chi : U \to B(M)$, defined by

$$\chi(n) = (n, D_{x_1}(n), \ldots, D_{x_d}(n)),$$

is horizontal.

Proof. Notice first that

$$H(d\chi \, D_{x_i}(n)) = E_i(\chi(n)).$$

Hence, if the image is horizontal we shall have on the image, and hence everywhere by equivariance, $[E_i, E_j] = 0$, which implies that curvature and torsion are zero, by the above lemma.

Conversely, if these forms vanish, then by theorem 5.5 there is a horizontal manifold N, and $E_1, \ldots, E_d$ are vector fields with trivial brackets. Pulling them down to M, we have by theorem 1.5 that there is a coordinate system $x_1, \ldots, x_d$ on M such that

$$D_{x_i} = d\pi \, E_i \mid_N$$

This coordinate system will then have the required property.

Problem 1. The property of having torsion zero is invariant under combinations of connexions. Using the existence proof in Chapter 5 show that every paracompact manifold has affine connexions with torsion zero.

6.1.5 Curvature and Torsion Transformations

The curvature and torsion forms give rise to tensors on M, which may be considered in the context of the tensor calculus on M, but we prefer to approach these concepts directly in terms of linear transformations on tangent spaces.

To every $m \in M$ and pair s, $t \in M_m$ we make correspond a linear transformation $R_{st}: M_m \to M_m$, called the *curvature transformation*, as follows. Let $b \in B(M)$ be such that $\pi(b) = m$, $\bar{t}$, $\bar{s} \in B(M)_b$ such that $d\pi(\bar{t}) = t$, $d\pi(\bar{s}) = s$. Then using the notation of 6.1.3, we define

$$R_{st}(u) = -b\Phi(\bar{s}, \bar{t}) \, b^{-1}(u)$$

for any $u \in M_m$.

Using the horizontality and equivariance of Φ it is easy to verify that the above definition is independent of the various choices made, namely, the choices of b, $\bar{s}$, and $\bar{t}$.

The definition of R_{st} may be rephrased in terms of matrices as follows. Let $b = (m, e_1, ..., e_d)$, $\bar{s}$, $\bar{t}$ be as before. Then R_{st} is the linear transformation of M_m whose matrix with respect to the basis $e_1, ..., e_d$ is $-\Phi(\bar{s}, \bar{t})$. For example, $R_{st}e_j = -\Sigma_i \Phi_{ij}(\bar{s}, \bar{t}) \, e_i$.

If X, Y, Z are vector fields on M, then we denote by $R_{XY}Z$ the vector field satisfying

$$R_{XY}Z(m) = R_{X(m)Y(m)} Z(m).$$

To each $m \in M$ and pair s, $t \in M_m$ we make correspond a tangent $T_{st} \in M_m$, called the *torsion translation*, as follows. Let $b \in B(M)$, $\bar{s}$, $\bar{t} \in B(M)_b$ be as in the above definition of the curvature transformation. Then define

$$T_{st} = -b\Omega(\bar{s}, \bar{t}).$$

The proof that T_{st} is well defined is similar to the proof for R_{st}. Also, for vector fields X and Y we have a vector field T_{XY}.

6.1.6 Geodesics

Let γ be C^∞ curve in M. A *vector field along* γ is a cross section of γ in $T(M)$, the tangent bundle of M. For example, γ_* defines a vector field along γ. A vector field X along γ is said to be a *parallel vector field along* γ if, for any u, v, $X(u)$ is the parallel translate of $X(v)$ along γ from $\gamma(u)$ to $\gamma(v)$. In terms of the connexion on the associated bundle $T(M)$ (see 5.4) this says that X defines a horizontal curve in $T(M)$.

A C^∞ curve γ in M is called a *geodesic* if its tangent vector field, γ_*, is parallel along γ.

Notice that a geodesic is a parametrized curve and not just a point set. And the only reparametrization of a geodesic which will again give a geodesic is a linear change in the parameter.

Theorem 2. If γ is a C^∞ curve in M, $\bar{\gamma}$ its horizontal lift through $b \in \pi^{-1}(\gamma(0))$ in $B(M)$, then γ is a geodesic if and only if there exists $c \in R^d$ such that $\bar{\gamma}$ is an integral curve of $E(c)$ if and only if $\omega(\bar{\gamma}_*)$ is constant, that is, when we express

$$\bar{\gamma}_*(u) = \sum f_i E_i(\bar{\gamma}(u))$$

the f_i are constant.

The proof of this is trivial. Applying the theorem that vector fields have unique integral curves we obtain:

Corollary. For every $m \in M$, $t \in M_m$ there exists a unique geodesic γ such that $\gamma(0) = m$ and $\gamma_*(0) = t$.

6.1.7 Geometric interpretation of R_{st} and T_{st}

In this section we relate R_{st} and T_{st} to parallel translation around an infinitesimal parallelogram with sides s and t.

We generate a family of "parallelograms" in a manner similar to the construction in theorem 1.4, except that we use geodesics instead of integral curves of vector fields. As we go around the broken geodesic starting at m, we carry s and t along by parallel translation, generating vector fields S and T along the curve. The pieces of the curve are then in turn, the geodesic with tangent vector field S, T, $-S$, and $-T$ and each is followed parameter distance u. The end point of this broken curve we call $\sigma(u)$, so σ is a C^∞ curve starting at m. Since the "parallelograms" are not generally closed, σ is not necessarily the constant curve. Let $A(u)$ be the linear transformation of M_m given by parallel translation along the broken curve and then backward along σ.

Theorem 3. The first order tangents of σ and A are 0. The second order tangent (see problem 1.20) of σ is $2T_{st}$, and $A''(0) = 2R_{st}$, where A is viewed as a curve in the vector space $\mathfrak{gl}(M_m)$.

In other words, the first order parts of the displacement given by traversing a parallelogram and parallel translation around such a parallelogram both vanish, and the second order parts are given by torsion and curvature, respectively.

Proof. If we lift the broken geodesic horizontally to $B(M)$ starting at b then we get integral curves of a pair of basic vector fields X and Y

in the way that we did in theorem 1.4. So by that theorem the curve γ generated by the end of these lifts, which is above σ, will have second order tangent $2[X, Y](b)$, with the first order tangent 0. Thus, by lemma 5 the horizontal component will be a lift of $2T_{st}$, which proves the first assertion.

Similarly, the vertical component of $2[X, Y](b)$ measures the second order deviation of γ from parallel translation, so A has second derivative equal to the corresponding transformation on M_m, which is $2R_{st}$ by lemma 5.

Development of curves in M into M_m. We compare curves in M and in flat affine space M_m by a process known as development. This is done by making the tangent vectors in each case have the same relation to the parallel translates of a particular basis.

Let γ be a broken C^∞ curve starting at m, $b = (m, e_1, ..., e_d)$ a basis at m. Let σ be the horizontal lift of γ to b in $B(M)$. Then we may write $\gamma_* = \sigma f$, where $f = \sigma^{-1}\gamma_*$ is a curve in R^d. Then $g(t) = \int_0^t f$ defines a curve in R^d which we call the development of γ into R^d with respect to b, and bg is a curve in M_m which we call the *development of γ into M_m*. bg is independent of the choice of basis b at m.

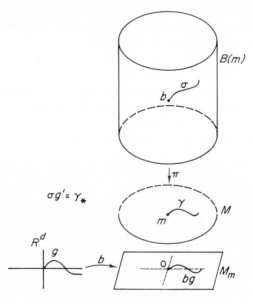

FIG. 21.

The process can be reversed; for every broken C^∞ curve τ in M_m starting at 0, there is a corresponding curve γ in M starting at m such that τ is the development of γ. To show this we get the horizontal lift σ of γ first; what we know is that $\sigma_*(t) = E(b^{-1}\tau'(t))(\sigma(t))$ so the existence then follows from:

Lemma 6.   Let E be a linear map from R^d into a linear space of C^∞ vector fields on a manifold N. Then for every broken C^∞ curve f in R^d and $n \in N$ there is a unique broken C^∞ curve σ in N such that

$$\sigma(0) = n \qquad \text{and} \qquad \sigma_*(t) = E(f(t))(\sigma(t)).$$

Proof.   Define vector field X on $U \times N$, U a neighborhood of 0 in R, by $X(t, n') = D_1(t) + E(f(t))(n')$. Then the integral curve of X starting at $(0, n)$ is the graph of σ, so σ exists.

Note that geodesics are developed into straight lines.

Problem 2.   Let $s, t \in M_m$ and let τ be a broken C^∞ closed curve in the plane of s, t such that $\tau(0) = 0$. For every $v \in R$ let $v\tau$ be the curve in M_m defined by $(v\tau)(u) = v(\tau(u))$. Let $\tau(u) = p(u) s + q(u) t$, τ parametrized on $[0, 1]$,

$$A = \int_0^1 p(u) \, q'(u) \, du,$$

the area enclosed by τ relative to s, t. Let h be the map of $[0, 1] \times R$ into M such that $h(\cdot, v)$ develops into $v\tau$. Let $\gamma(v) = h(1, v)$, and let $S(v)$ be the linear transformation of M_m given by parallel translation around the closed curve consisting of $h(\cdot, v)$ and a piece of γ^{-1}. Prove the following generalization of theorem 3:

$\gamma_*(0) = 0$ and the second order tangent is $2AT_{st}$.

$S'(0) = 0$ and $S''(0) = 2AR_{st}$.

6.2 The Structural Equations of an Affine Connexion

Let H be an affine connexion on $B(M)$. Let ϕ, ω, Φ, Ω be the 1-form of the connexion, the solder form, the curvature form, and the torsion form, respectively.

Theorem 4.   We have

$$d\omega = -\phi\omega + \Omega,$$
$$d\phi = -\tfrac{1}{2}[\phi, \phi] + \Phi$$

These are called the first and second structural equations of the connexion. (See 4.9 for definitions of the forms $\phi\omega$, $[\phi, \phi]$.)

Proof.   The second structural equation is simply the structural equation for a connexion on a principal bundle (see theorem 5.3). To prove the first structural equation we prove the following more general result.

Theorem 5.   Let θ be an R^d-valued equivariant horizontal p-form on $B(M)$. Then

$$d\theta = -\phi\theta + D\theta.$$

Proof.   The proof is almost identical to that of theorem 5.6. We proceed by evaluating each side on vector fields $Y_1, ..., Y_{p+1}$ which are drawn from a collection of vector fields which locally span the tangent space to $B(M)$. We consider the same cases as before.

(i) *No Y_i is vertical.* Hence we may assume all Y_i are horizontal. But then the $\phi\theta$ term is 0, and the remainder is merely the definition of $D\theta$.

(ii) *One Y_i is vertical.* Assume $Y_{p+1} = \lambda X$, $X \in \mathfrak{gl}(d, R)$. As before we choose right invariant horizontal Y_i's such that $[Y_i, \lambda X] = 0$, $i = 1, ..., p$. Then by theorem 4.2,

$$d\theta(Y_1, ..., Y_{p+1}) = (-1)^p (\lambda X) \, \theta(Y_1, ..., Y_p).$$

Now applying the lemma 5.3 with the representation $\phi = $ matrix operation on R^d, we have

$$(-1)^p \lambda X \theta(Y_1, ..., Y_p) = (-1)^{p+1} X\theta(Y_1, ..., Y_p).$$

On the other hand, $D\theta(Y_1, ..., Y_p, \lambda X) = 0$, since $D\theta$ is horizontal, so the right-hand side gives

$$-\phi\theta(Y_1, ..., Y_p, \lambda X) = -\sum_{i=1}^{p+1} (-1)^{i-1} \phi(Y_i)\theta(Y_1, ..., Y_{i-1}, Y_{i+1}, ..., Y_{p+1})$$
$$= -(-1)^p \phi(\lambda X) \, \theta(Y_1, ..., Y_p)$$
$$= (-1)^{p+1} X\theta(Y_1, ..., Y_p).$$

We have therefore reduced both sides to the same function, as desired.

(iii) *Two of the Y_i are vertical.* In this case everything is 0, and the equality holds automatically. QED

Theorem 6. Let θ be an equivariant horizontal p-form with values in either $\mathfrak{gl}(d, R)$ or R^d. Let $\square$ denote the operation of $\mathfrak{gl}(d, R)$ on either $\mathfrak{gl}(d, R)$ or R^d by bracketing or matrix multiplication, respectively. Then

$$D^2\theta = \Phi \,\square\, \theta.$$

Proof. θ satisfies the structural equation

$$d\theta = -\phi \,\square\, \theta + D\theta,$$

either by theorem 5.6 or theorem 5 above. Applying D to both sides yields

$$d^2\theta \circ H = -d(\phi \,\square\, \theta) \circ H + D^2\theta.$$

But $d^2 = 0$, and

$$d(\phi \,\square\, \theta) \circ H = d\phi \circ H \,\square\, \theta - \phi \circ H \,\square\, D\theta$$
$$= \Phi \,\square\, \theta.$$

and this gives the result.

Combining this with the Bianchi identity (theorem 5.4) gives the *affine Bianchi indentities*: $D\Phi = 0$, $D\Omega = \Phi\omega$.

6.2.1 Dual Formulation of the Structural Equations

The first structural equation has a dual in terms of the bracket of a fundamental and a basic vector field.

Theorem 7. Let $X \in \mathfrak{gl}(d, R)$, $x \in R^d$. Then

$$[\lambda X, E(x)] = E(Xx),$$

where Xx is the action of the matrix X on the vector x.

Proof. We can prove that $[\lambda X, E(x)]$ is horizontal by means of the second structural equation or notice that this is precisely what the lemma 5.2 says. Thus we need only show that

$$\omega([\lambda X, E(x)]) = Xx.$$

But by the first structural equation

$$\omega([\lambda X, E(x)]) = +\lambda X \omega(E(x)) - E(x)\,\omega(\lambda X) - d\omega(\lambda X, E(x)) \qquad (4.6.)$$

$$= \phi\omega(\lambda X, E(x))$$

$$= \phi(\lambda X)\,\omega(E(x))$$

$$= Xx. \qquad \text{QED}$$

For the vector fields E_i, E_{jk} this theorem gives the formula $[E_{jk}, E_i] = \delta_{ik}E_j$.
We remark that this formula also follows directly from theorem 3.1 and hence can in turn be used to prove the first structural equation.

Problem 3.   Give an alternate proof of theorem 7 by applying the formula $L_{\lambda X} = i(\lambda X)\,d + di(\lambda X)$ to the solder form ω and evaluating on the vector field $E(x)$. (Compare with problem 5.4.)

6.2.2 Difference Forms

Let ϕ, ψ be two connexion 1-forms on $B(M)$. Then we define the *difference form* τ by $\tau = \psi - \phi$.

Lemma 7.   The difference form τ has the following properties:

(i) $\tau \in C^\infty$,

(ii) τ is horizontal,

(iii) τ is equivariant.

Conversely, if ϕ is a connexion form and τ is a $\mathfrak{gl}(d, R)$-valued 1-form satisfying (i)-(iii), then $\phi + \tau$ is again the 1-form of a connexion.

These facts are immediate.

A difference form τ gives rise to a linear transformation $T_s : M_m \to M_m$ for each $s \in M_m$; namely, if $\pi(b) = m$ we define

$$T_s t = b\tau(\bar{s})\,b^{-1}t,$$

where $\bar{s} \in B(M)_b$ is such that $d\pi\,\bar{s} = s$.

Conversely, given a function T which assigns to each $m \in M$, $s \in M_m$ a linear transformation $T_s : M_m \to M_m$ such that T is linear in s and "differentiable" in m and s [that is, as a function from $T(M)$ to the bundle of linear transformations of the tangent spaces to M, which is the bundle with fibre $\mathfrak{gl}(d, R)$ and action the adjoint representation of $Gl(d, R)$ associated to $B(M)$], then we can define a $\mathfrak{gl}(d, R)$-valued 1-form satisfying (i)-(iii) by

$$\tau(\bar{s}) = b^{-1}T_{d\pi\bar{s}}b$$

for any $\bar{s} \in B(M)_b$. Also $T_s t = b\tau(\bar{s})\, b^{-1}t$ holds, which shows there is a one-to-one correspondence between difference forms and certain functions which we shall call *linear transformation fields*.

Theorem 8.   Let ϕ be a connexion on $B(M)$. Then there is a difference form τ such that $\psi = \phi + \tau$ is the 1-form of a connexion whose torsion 2-form is identically zero (cf. problem 1).

Further, if τ' also has this property, then $\tau\omega = \tau'\omega$ (see 4.9 for notation).

Proof.   We assume τ exists and see what conditions it must satisfy. Writing down the first structural equations for ϕ and ψ, we have

$$d\omega = -\phi\omega + \Omega$$
$$d\omega = -\psi\omega,$$

since ψ has torsion zero. So $(\phi - \psi)\,\omega = \Omega$, and hence τ must satisfy the equation

$$-\tau\omega = \Omega.$$

The last part of the theorem is now immediate. To show that such a τ exists, we simply define, for $b \in B(M)$, $s \in B(M)_b$, $x \in R^d$,

$$-\tau(s)(x) = \tfrac{1}{2}\Omega(s, E(x)(b)).$$

It is not hard to verify that τ is a C^∞, equivariant, horizontal 1-form. Further, for $b \in B(M)$, $s, t \in B(M)_b$,

$$-\tau\omega(s, t) = -\tau(s)\,\omega(t) + \tau(t)\,\omega(s)$$
$$= \tfrac{1}{2}\Omega(s, E(\omega(t))(b)) - \tfrac{1}{2}\Omega(t, E(\omega(s))(b))$$
$$= \tfrac{1}{2}\Omega(s, t) - \tfrac{1}{2}\Omega(t, s)$$
$$= \Omega(s, t),$$

as desired.

Notice that the linear transformation field T associated with τ is $T_s(t) = \tfrac{1}{2}T_{st}$, where T_{st} is the torsion translation corresponding to s, $t \in M_m$ (6.1.5).

Before proceeding we give a convenient formulation of parallel translation.

Lemma 8. Let ρ be any C^∞ curve in M, $t \in M_{\rho(0)}$, $b \in B(M)$ such that $\pi(b) = \rho(0)$. Let ϕ be a connexion form on $B(M)$ and β the ϕ-horizontal lifting of ρ through b. Then the ϕ-parallel translate of t along ρ to $\rho(u)$ is $\beta(u)\, b^{-1}t$.

The proof is immediate in view of the action of any $c \in B(M)$ in mapping R^d isomorphically onto $M_{\pi(c)}$.

The following is a useful characterization of geodesics.

Lemma 9. Let γ be a curve in M, $\bar{\gamma}$ a not necessarily horizontal lift of γ to $B(M)$. Then γ is a geodesic if and only if

$$(\bar{\gamma}_* + \phi(\bar{\gamma}_*))\, \omega(\bar{\gamma}_*) = 0.$$

Proof. Let P be the principal bundle induced by γ over the domain U of γ from $B(M)$. Thus

$$P = \{(u, b) \mid u \in U, \pi b = \gamma(u)\}.$$

Then the map $\alpha : P \to B(M)$, $\alpha(u, b) = b$ is a C^∞ map, so we may pull back the solder and connexion forms, $\omega' = \alpha^*\omega$, $\phi' = \alpha^*\phi$, $\Omega' = \alpha^*\Omega = 0$ since the horizontal space of P is one-dimensional.

$\sigma : U \to P$ given by $\sigma(u) = (u, \bar{\gamma}(u))$ is a C^∞ curve in P, and σ_* extends to a right invariant C^∞ vector field X on P. If we do the same thing using a horizontal lift β of γ instead of $\bar{\gamma}$, we get a C^∞ vector field Y on P such that $\phi'(Y) = 0$. Moreover, X and Y both project to $D = d/du$ on U, so $[X, Y]$ is vertical and $\omega'(X) = \omega'(Y)$.

By theorem 2, γ is a geodesic if and only if $\omega(\beta_*)$ is constant, or, $\beta_* \omega(\beta_*) = 0$, which by applying α becomes

$$Y\omega'(Y) = 0.$$

Now the first structural equation gives

$$d\omega'(X, Y) = X\omega'(Y) - Y\omega'(X)$$

$$= -\phi'(X)\, \omega'(Y), \quad \text{or, since} \quad \omega'(X) = \omega'(Y),$$

$$X\omega'(X) + \phi'(X)\, \omega'(X) = Y\omega'(Y).$$

Thus the condition for γ to be a geodesic is that the left side of this equation is 0, and this becomes, along points of σ,

$$\frac{d}{du}\, \omega'(\sigma_*) + \phi'(\sigma_*)\, \omega'(\sigma_*) = 0.$$

Inserting the meaning of ω', ϕ' we have

$$\frac{d}{du}\, \omega(\bar{\gamma}_*) + \phi(\bar{\gamma}_*)\, \omega(\bar{\gamma}_*) = 0.$$

This is the conclusion desired in one direction. To go the other way we only need reverse the implication

$$X\omega'(X) + \phi'(X)\, \omega'(X) = 0$$
$$\Rightarrow \frac{d}{du}\, \omega'(\sigma_*) + \phi'(\sigma_*)\, \omega'(\sigma_*) = 0;$$

this is possible by the right invariance of X.

The following theorem is not relevant to Riemannian geometry, but has interesting consequences for the theory of connexions.

Theorem 9.   Let ϕ and ψ be two connexion forms on $B(M)$, τ their difference form. Then these two connexions have the same geodesics in M if and only if for every $b \in B(M)$, $s \in B(M)_b$,

$$\tau(s)\, \omega(s) = 0.$$

Proof.   If the two connexions have the same geodesics, then given $s \in B(M)_b$ there is a geodesic γ such that a lift $\bar{\gamma}$ of γ satisfies $\bar{\gamma}_*(0) = s$. (If s is vertical $\gamma =$ constant.)

Then by lemma 9

$$\bar{\gamma}_*(0)\, \omega(\bar{\gamma}_*) = -\phi(s)\, \omega(s)$$
$$= -\psi(s)\, \omega(s) \qquad \text{as desired.}$$

Conversely, if $\tau(s)\, \omega(s) = 0$ for all s, then if $\bar{\gamma}$ is a lift of a ϕ-geodesic γ,

$$\bar{\gamma}_*\omega(\bar{\gamma}_*) = -\phi(\bar{\gamma}_*)\, \omega(\bar{\gamma}_*)$$
$$= -\psi(\bar{\gamma}_*)\, \omega(\bar{\gamma}_*), \qquad \text{since} \qquad \tau(\bar{\gamma}_*)\, \omega(\bar{\gamma}_*) = 0,$$

so γ is a ψ-geodesic.

The following is a generalization of theorem 8 and a comparison of the proofs shows that the two connexions occurring there have the same geodesics.

Corollary. Let ϕ be a connexion and θ an equivariant horizontal R^d-valued 2-form on $B(M)$. Then there is a unique connexion ψ having θ as its torsion form and the same geodesics as ϕ.

In particular, if the geodesics and torsion forms of two connexions coincide, so do the connexions.

Proof. The proof of uniqueness will give a formula for $\tau = \phi - \psi$, hence existence will follow from the trivial verification that τ is horizontal and equivariant.

So let us suppose that ϕ and ψ have the same geodesics and that the first structural equations are

$$d\omega = -\phi\omega + \Omega$$
$$d\omega = -\psi\omega + \theta.$$

Subtracting these gives

$$\tau\omega = (\phi - \psi)\,\omega = \Omega - \theta = \eta,$$

that is, for $s, t \in B(M)_b$,

$$\tau(s)\,\omega(t) - \tau(t)\,\omega(s) = \eta(s, t).$$

By polarization, the condition for equality of geodesics becomes $\tau(s)\,\omega(t) + \tau(t)\,\omega(s) = 0$. Adding the last two equations gives a formula for τ, since $\omega(t)$ is arbitrary:

$$\tau(s)\,\omega(t) = \tfrac{1}{2}\eta(s, t). \qquad \text{QED}$$

Problem 4. Show that if two connexions have the same geodesics then so does any combination of them in the sense of problem 5.2.

Problem 5. *Connexions on parallelizable manifolds.* Let ρ be an R^d-valued 1-form on M which gives a parallelization of M. We define three connexions associated with ρ as follows.

The *direct connexion* is the one for which the vector fields $\rho^{-1}(x)$, for a fixed $x \in R^d$, are parallel along every curve. The *torsion zero connexion* is the one with the same geodesics as the direct connexion but with torsion zero. The *opposite connexion* is the one with connexion form $\phi + 2\tau$, where ϕ is the connexion form of the direct connexion and $\phi + \tau$ is the form of the torsion zero connexion.

Prove the following facts about these connexions:

(a) The parallelization ρ leads naturally to a cross section of M in $B(M)$. This cross section is horizontal with respect to the direct connexion and by means of it the structural equations pull down to one equation on $M : d\rho = P$, where P is the torsion form pulled down.

(b) Two parallelizations are related by a C^∞ function from the manifold into $Gl(d, R)$ and the two corresponding direct connexions are the same if and only if the function is a constant (M connected).

(c) The geodesics of all three connexions are the same and consist of the integral curves of the vector fields $\rho^{-1}(x)$.

(d) If ρ is constant on vector fields X, Y, Z, then the torsion and curvature of these connexions are given by the table:

direct connexion	$T_{XY} = [X, Y]$,	$R_{XY} = 0$
torsion zero connexion	$T_{XY} = 0$,	$R_{XY}Z = \frac{1}{4}[[X,Y],Z]$
opposite connexion	$T_{XY} = -[X, Y]$,	$R_{XY} = 0$.

(e) For the opposite connexion $R_{XY} = 0$ if and only if ρ is constant on $[X, Y]$.

If ρ is constant on all such $[X, Y]$ then it is well known that a local Lie group structure can be given M such that the constant fields are the left invariant vector fields. More generally, the problem of local equivalence of direct connexions has been studied in the larger context of G-structures. A lucid account of this matter may be found in a book by Shlomo Sternberg, "Lectures on Differential Geometry," Prentice-Hall, Inc., Englewood Cliffs, New Jersey, 1964.

Problem 6. *Connexions on Lie groups.* A Lie group has a parallelization by means of left invariant vector fields and also by right invariant vector fields. We call the corresponding direct connexions the *left connexion* and the *right connexion*. Show that they are opposite connexions of each other, so the two torsion free connexions coincide. The first structural equation of the left connexion pulls down, as in problem 5(a), to the equation of Maurer-Cartan (problems 4.20, 4.23). Hence express P in terms of the structural constants c_{ij}^k.

The three connexions coincide if and only if the group is Abelian, so the various products of Euclidean space and tori have flat affine connexions with torsion zero.

6.3 The Exponential Maps

The *exponential map* at $m \in M$ is a mapping of a neighborhood U of $0 \in M_m$ into M,

$$\exp_m : U \to M.$$

For those $t \in M_m$ for which $\exp_m(t)$ is defined, it is given as follows. Let γ be the unique geodesic in M such that $\gamma(0) = m$ and $\gamma_*(0) = t$. Then define

$$\exp_m(t) = \gamma(1).$$

Note that $\exp_m(ut) = \gamma(u)$, u a real number, if $\gamma(u)$ exists. The domain of $\exp_m$ is an open set in M_m which is star-shaped with respect to $0 \in M_m$ in the sense that if t is in the domain then so is the line segment from 0 to t.

Besides this exponential map we also shall consider a certain lifting to $B(M)$. For $b \in B(M)$, with $\pi(b) = m$, we define $\overline{\exp}_b(t) = \bar{\gamma}(1)$, where $\bar{\gamma}$ is the unique horizontal lift of γ through b. Since γ is a geodesic, $\bar{\gamma}$ is an integral curve of $E(x)$, where $bx = \gamma_*(0) = t$ (theorem 2).

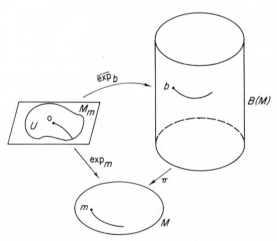

FIG. 22.

We will show that $\overline{\exp}_b \in C^\infty$, and hence also $\exp_m = \pi \circ \overline{\exp}_b \in C^\infty$. It will immediately follow that $\exp_m$ is a diffeomorphism onto a neighborhood of m, since $d \exp_m$ maps onto M_m; in fact, if $e_1, ..., e_d$ is a basis for M_m, $u_1, ..., u_d$ the dual base, then $d \exp_m D_{u_i}(0) = e_i$, and the dimensions of M_m and M are the same.

Theorem 10. $\overline{\exp}_b \in C^\infty$.

Proof.   We proceed in a slightly more general fashion. Consider the *affine bundle* $A(M)$ over M, that is, the bundle with bundle space consisting of pairs (b, t), $b \in B(M)$, $t \in M_{\pi(b)}$, and projection map $(b, t) \to \pi(b)$. This is a manifold under the obvious definition of the differentiable structure. We define a mapping

$$F : B(M) \times A(M) \to T(B(M))$$

by

$$F(b, c, t) = E(c^{-1}t)\,(b).$$

For each $(c, t) \in A(M)$ we have a vector field $E(c^{-1}t)$ on $B(M)$. F is clearly a C^∞ mapping. By the theorems on differential equations in the appendix, there is a C^∞ mapping G of a neighborhood of $\{0\} \times B(M) \times A(M)$ into $B(M)$ given by $G(u, b, c, t) = \gamma(u)$, where γ is the integral curve of $E(c^{-1}t)$ with $\gamma(0) = b$. Then

$$\overline{\exp}_b t = G(1, b, b, t). \qquad \text{QED}$$

Corollary.   The map Exp: $T(M) \to M$, defined by

$$\text{Exp}(m, t) = \exp_m t,$$

is defined on a neighborhood of the trivial cross section of $T(M)$ and is C^∞ there.

Proof.   For $m \in M$ choose a C^∞ cross section χ of a neighborhood of m into $B(M)$. Then on this neighborhood the mapping

$$J : (m, t) \to (1, \chi(m), \chi(m), t)$$

into $\{u\} \times B(M) \times A(M)$ is C^∞, and we have

$$\text{Exp} = \pi \circ G \circ J. \qquad \text{QED}$$

Problem 7. (a)   If P is a principal bundle over manifold M, M having an affine connexion, $m \in M$, then $\exp_m$ may be factored through P.
(b)   Every bundle over R^d is trivial.

6.3.1 Completeness

An affine connexion is called *complete* if all geodesics can be infinitely extended, that is, if each exponential map is defined on the whole tangent space. This is equivalent to saying that the local group of transformations of $B(M)$ generated by any basic vector field $E(x)$ can be extended to a global one-parameter group of transformation of $B(M)$. We shall see in Chapter 8 that completeness of a Riemannian connexion is equivalent to the completeness of the Riemannian metric.

6.3.2 Normal Coordinates

A coordinate map $\phi : U \to R^d$, $U \subset M$, is called a *normal coordinate map* at $m = \phi^{-1}(0)$ if the pre-images of rays through $0 \in R^d$ are geodesics, where a ray is a straight line of the form $u \to ux$, $x \in R^d$.
If we choose a basis $b \in B(M)$ with $\pi(b) = m$, then we have an identification of R^d with M_m. Combining this with $\exp_m$ and applying the theorem we see that the function $\exp_m \circ b$ is the inverse of a normal coordinate system at m.
A *normal coordinate neighborhood* N, the domain of a normal coordinate map ϕ, has the property that every $n \in N$ can be joined to $\phi^{-1}(0)$ by a unique geodesic in N.
We remark that if the curvature and torsion of our connexion both vanish, then by theorem 1 there are coordinate systems whose inverses send any lines in $U \subset R^d$ into geodesics, that is, the coordinate system is normal with respect to each of its points. This is the affine version of the local isometry of a flat Riemannian manifold with Euclidean space (see corollary, theorem 9.3).

Problem 8.   Show that the exponential map at the identity of a Lie group, for any of the connexions of problem 6, is the same as the exponential map of the Lie group when we identify the Lie algebra with the tangent space at the identity, so these connexions are all complete.
Hence, in view of problem 2.2, the exponential map of a complete affine connexion need not be onto even though the manifold be connected.

Problem 9.   If $G = S^3$, as a subgroup of the multiplicative quaternions, show that the geodesics of S^3 are great circles.

6.4 Covariant Differentiation and Classical Forms

6.4.1 Covariant Derivatives

We can use parallel translation in $T(M)$ to give a notion of directional derivatives of vector fields. More generally, this can be done in any vector bundle associated with $B(M)$, so that each representation of $Gl(d, R)$ on a vector space F gives rise to the concept of covariant differentiation of cross sections of the bundle with fibre F. We give several definitions of this notion, ignoring the problems of equivalence and independence of the choice of curve. We fix an affine connexion H with form ϕ.

Let (W, F, G, M) be a vector bundle, associated with $B(M)$, with fibre F, and group $G = Gl(d, R)$. Then each $b \in B(M)$ gives an isomorphism of F onto the fibre of W over $\pi(b)$ such that for $f \in F$, $g \in G$ we have $b(gf) = (bg)f$ (3.3). Let $U \subset M$ be a neighborhood of $m \in M$ and let $X : U \to W$ be a cross section of U. Finally take $t \in M_m$. We shall give several definitions of $\nabla_t X$, the *covariant derivative of X with respect to t*. The notation $D_t X$ is also commonly used. In the first place, $\nabla_t X$ will be an element of the fibre of W over m. If Y is a vector field on U, then $\nabla_Y X$ will denote the cross section over U given by $\nabla_Y X(n) = \nabla_{Y(n)} X$, the covariant derivative of X in the direction $Y(n)$.

(i) $\nabla_t X$ gives a measure of how much X fails to be horizontal in the direction t. Thus it is a comparsion between the lifting of t given by X, namely $dX\,t$, and the horizontal lifting of t, $H'(dX\,t)$, where H' is the connexion on W arising from the affine connexion on $B(M)$ (5.4). Since the fibres of W are vector spaces we may identify the vertical tangents with elements of the fibre, in the usual way in which a vector space is identified with its tangent space at any point (2.5). Using this identification we define

$$\nabla_t X = V(dX\,t) = dX\,t - H'(dX\,t).$$

(ii) $\nabla_t X$ is differentiation with respect to parallel translation. Let γ be a curve with $\gamma_*(0) = t$, let $e_1(u), ..., e_k(u)$ be a basis of the fibre of F over $\gamma(u)$ such that each e_i is a horizontal lift of γ, that is, $e_i(u)$

is obtained by parallel translation of $e_i(0)$ along γ to $\gamma(u)$ (5.4). Define real-valued, C^∞ functions f_i by $X(\gamma(u)) = \Sigma_i f_i(u) e_i(u)$. Then we have

$$\nabla_t X = \sum_i f_i'(0) e_i(0).$$

(iii) $\nabla_t X$ corresponds to the derivative in a horizontal direction of a function on $B(M)$ associated to X.

For each $b \in B(M) \cap \pi^{-1}(U)$ we define

$$\tilde{X}(b) = b^{-1}X(\pi b),$$

so that $\tilde{X}$ is a function from $B(M) \cap \pi^{-1}(U)$ into F. Note that $\tilde{X}(bg) = g^{-1}b^{-1}X(\pi(b)) = g^{-1}\tilde{X}(b)$. Conversely, any C^∞ function $\tilde{X} : B(M) \to F$ such that $\tilde{X}(bg) = g^{-1}\tilde{X}(b)$ gives rise to a cross section $X : M \to W$, namely, $\pi(b) \to b\tilde{X}(b)$.

In this case it is just as easy to define $\nabla_Y X$ as $\nabla_t X$. Let $\bar{Y}$ be the unique horizontal lift of Y to W, so that $\bar{Y}(b)$ is the unique horizontal tangent such that $d\pi \, \bar{Y}(b) = Y(\pi(b))$. Then we have

$$\nabla_Y X = \text{the cross section associated with the function } \bar{Y}\tilde{X}.$$

Problem 10. Show that the definitions of covariant derivative given are all the same.

We give some examples.

(i) If $F = R^d$, $W = T(M)$, then X is just a vector field. If X is parallel along γ then $\nabla_{\gamma_*} X = 0$, and conversely.

Note that in this case, the $\tilde{X}$ of (iii) above is just $\omega(\bar{X})$, where $\bar{X}$ is any lift of X to $B(M)$. Using this and the structural equations allows us to get a convenient formulation for $\nabla_t X$.

Theorem 11. Let X be a vector field on M, $t \in M_m$, $\bar{t}$ a lift of t to $b \in B(M)$, $\bar{X}$ a lift of X.
Then $\nabla_t X = b(\bar{t}\omega(\bar{X}) + \phi(\bar{t}) \, \omega(\bar{X}(b)))$.

Proof. Let Y be an extension of $\bar{t}$ to a right invariant vector field on $B(M)$. Then $[Y - HY, \bar{X}]$ is vertical, being π-related to $[0, X]$, so

$$
\begin{aligned}
d\omega(Y - HY, \bar{X}) &= (Y - HY)\,\omega(\bar{X}) - \bar{X}\omega(Y - HY) - \omega([Y - HY, \bar{X}]) \\
&= (Y - HY)\,\omega(\bar{X}) \\
&= -\phi(Y - HY)\,\omega(\bar{X}) + \phi(\bar{X})\,\omega(Y - HY) \\
&= -\phi(Y)\,\omega(\tilde{X}).
\end{aligned}
$$

Evaluating at b we get

$$H\bar{t}\omega(\bar{X}) = \bar{t}\omega(\bar{X}) + \phi(\bar{t})\,\omega(\bar{X}(b)).$$

Applying b to $H\bar{t}\omega(\bar{X})$ gives $\nabla_t X$ by (iii), which proves the formula. Comparing with the result of lemma 9 easily yields:

Corollary. γ is a geodesic if and only if $\nabla_{\gamma_*}\gamma_* = 0$.

   (ii) If $F = \mathfrak{gl}(d, R)$ and the action of G on F is simply the adjoint representation, then W is the bundle which has fibre above m equal to the set of all linear transformations of M_m into itself. The cross sections correspond to horizontal, equivariant, R^d-valued 1-forms on $B(M)$.

   (iii) G acts on R^{d*} as follows (4.5). If $f \in R^{d*}$, $g \in G$, $v \in R^d$, then

$$(gf)(v) = f(g^{-1}v).$$

The corresponding bundle W is just the Grassmann bundle G^1, and cross sections are 1-forms. The action of G on R^{d*} can be extended uniquely to a group of homomorphisms of the Grassmann algebra over R^{d*}. This action leads to covariant differentiation of differential forms.

   (iv) By combining the actions of G on R^d and R^{d*}, we get an action on tensor products of copies of R^d and R^{d*}. The cross sections of the corresponding associated vector bundles are tensor fields. If there are r copies of R^d and s copies of R^{d*}, then the tensor fields are said to be of type (r, s). For example, the torsion translation is a tensor field of type $(1, 2)$, while the curvature transformation is a tensor of type $(1, 3)$. Example (ii) above is actually the case of tensors of type $(1, 1)$.

   (v) If X is a cross section into W and f a real-valued function on M, then fX is again a cross-section over M. Its covariant derivative is related to that of X as follows. Let $t \in M_m$, then

$$\nabla_t(fX) = f(m)\nabla_t X + (tf)\,X(m).$$

Proof.   By the third definition of covariant derivative, we have that

$$\widetilde{fX}(b) = b^{-1}(f(\pi b)\,X(\pi b))$$

$$= f(\pi b)\,b^{-1}X(\pi b)$$

$$= f(\pi b)\,\tilde{X}(b).$$

That is

$$\widetilde{fX} = (f \circ \pi)\check{X}.$$

Hence

$$\bar{\imath}(\widetilde{fX}) = \bar{\imath}(f \circ \pi)\,\check{X}(b) + f \circ \pi(b)\,\bar{\imath}\check{X}$$

$$= (tf)\,\check{X}(b) + f(\pi b)\,\bar{\imath}\check{X},$$

and applying b to both sides yields

$$\nabla_t(fX) = (tf)\,X(\pi b) + f(\pi b)\,\nabla_t X.$$

Problem 11. For parallelizable M show that cross sections of a tensor bundle are in one-to-one correspondence with functions of M into the fibre F and that covariant differentiation for the direct connexion corresponds to differentiation of these vector-valued functions.

Problem 12. For parallelizable M and vector fields X, Y as in problem 5(d) derive the following formulas for covariant derivatives:

direct connexion	$\nabla_X Y = 0$
torsion zero connexion	$\nabla_X Y = \frac{1}{2}[X,\ Y]$
opposite connexion	$\nabla_X Y = [X,\ Y].$

Problem 13. Let M be a d-dimensional manifold and let ∇ be the covariant derivative of a torsion zero affine connexion over M. Let θ be a differential form on M, and let X_1, ..., X_d be a parallelization of a neighborhood U of M, β_1, ..., β_d the dual 1-forms on U. Prove the following formula for the exterior derivative of θ in U:

$$d\theta = \sum_i \beta_i \nabla_{X_i} \theta.$$

(*Hint*: Show that the operator on the right is an antiderivation and that it agrees with d on functions and 1-forms.)

Hence conclude that a differential form is closed if there exists a torsion zero affine connexion with respect to which covariant derivatives of the form are zero; that is, the form is parallel along every curve. (A differential form θ is *closed* if $d\theta = 0$.)

Problem 14. Let θ be a 2-form and X, Y, Z vector fields. Show that

$$(\nabla_X \theta)(Y, Z) = X\theta(Y, Z) - \theta(\nabla_X Y, Z) - \theta(Y, \nabla_X Z).$$

Generalize this.

6.4.2 Covariant Derivative Definition of a Connexion

Covariant differentiation in the tangent bundle is sufficient to determine the connexion, since by the first example above it gives a differential equation $\nabla_{\gamma_*} X = 0$ for parallel translation along γ. This differential equation is linear and hence has a unique solution for any initial condition.

Thus an affine connexion is specified as soon as we give for each $m \in M$, $t \in M_m$, and vector field X, an element $\nabla_t X \in M_m$ such that

(i) $\nabla_t X$ is linear in t,

$$\nabla_{ut+vs} X = u(\nabla_t X) + v(\nabla_s X),$$

for $u, v \in R$ and $t, s \in M_m$;

(ii) if f is a real-valued, C^∞ function on M, then

$$\nabla_t(fX) = (tf)\, X(m) + f(m)\, \nabla_t X.$$

Sometimes it is convenient to turn the covariant derivative around, that is, for each C^∞ vector field X defined on an open set U of M we consider linear transformation $\tau(X)$, defined on each M_m with $m \in U$, by the formula $\tau(X)\, t = \nabla_t X$. A connexion is then given by hypothesizing the existence of such $\tau(X)$'s which satisfy the following condition, corresponding to (ii): $\tau(fX)\, t = f(m)\, \tau(X)\, t + (tf)\, X(m)$ (see [65]).

We now give a direct relation between covariant differentiation of vector fields and the connexion form on $B(M)$. It will depend on a local cross section of M in $B(M)$. Let $X_1, \ldots, X_d$ be vector fields defined on an open set $U \subset M$ such that

$$\chi : m \to (m, X_1(m), \ldots, X_d(m))$$

is a cross section $U \to B(M)$. For any vector fields $Y = \Sigma_i f_i X_i$ and X on U we define

$$L(XY) = \sum_i (Xf_i)\, X_i.$$

Then the relation between the covariant derivative of Y in the direction of X and the connexion form ϕ is

(iii) $\nabla_{X(m)} Y = \chi(m)\, \phi(d\chi\, X(m))\, \chi(m)^{-1} Y(m) + L(XY)(m),$

where $\chi(m)$ is now considered as a mapping of R^d to M_m.

Since ϕ is known on vertical tangents and the vertical tangents along with those of the form $d\chi t$ span $B(M)_{\chi(m)}$, formula (iii) is enough to define ϕ on all of $B(M)_{\chi(m)}$, and then by equivariance, on all of $\pi^{-1}(U)$.

Problem 15. Show that covariant derivatives of connexions combine in the same way as the connexions, that is, if ∇ is covariant derivative for connexion H, Γ is covariant derivative for connexion K, f a C^∞ function on M, then the covariant derivative for $(f \circ \pi)H + (1 - f \circ \pi)K$ is $f\nabla + (1 - f)\,\Gamma$.

6.4.3 The Structural Equations [66]

The structural equations correspond to formulas for the torsion and curvature transformations in terms of covariant derivatives. The formulas are:

(i) $$T_{XY} = [X, Y] - \nabla_X Y + \nabla_Y X,$$

(ii) $$R_{XY} = \nabla_{[X, Y]} - \nabla_X \nabla_Y + \nabla_Y \nabla_X.$$

The first formula informs us that torsion tells how much $\nabla_X Y - \nabla_Y X$ fails to be $[X, Y]$, and the second that curvature tells how much ∇ fails to be a Lie algebra homomorphism.

Problem 16. Prove (i) and (ii).

6.4.4 Coordinates

In any given coordinate system the connexion in terms of covariant derivatives can be expressed in terms of sufficiently many real-valued functions on M, which are indexed so as to make manipulations easier. These functions are called the *coefficients of the connexion*. Other functions give the coefficients of the torsion translation, the curvature transformation, and their covariant derivatives. The way in which these coefficients change when the coordinate system on M is changed can be found by using the chain rule for partial derivatives and the properties of the covariant derivative.

We define these classical coefficient functions. Let $x_1, \ldots, x_d$ be the coordinate system on M, and put $X_i = D_{x_i}$.

Then we have

$$\nabla_{X_i} X_j = \sum_{k=1}^{d} \Gamma_{ij}{}^k X_k ,$$

$$T_{X_i X_j} = \sum_k T_{ji}{}^k X_k ,$$

$$R_{X_j X_k}(X_m) = - \sum_i R^i{}_{mjk} X_i ,$$

$\nabla R(X_j , X_k , X_s)(X_m)$ (by definition)

$$= \nabla_{X_s}(R_{X_j X_k}(X_m)) - R_{X_j X_k}(\nabla_{X_s} X_m)$$

$$- R_{\nabla_{X_s} X_j X_k}(X_m) - R_{X_j \nabla_{X_s} X_k}(X_m)$$

$$= - \sum_i R^i{}_{mjk,s} X_i.$$

As a consequence of the equations of the last section we have

$$T_{ij}{}^k = \Gamma_{ij}{}^k - \Gamma_{ji}{}^k,$$

and

$$R^i{}_{mjk} = X_j \Gamma_{km}{}^i - X_k \Gamma_{jm}{}^i$$

$$+ \sum_s (\Gamma_{js}{}^i \Gamma_{km}{}^s - \Gamma_{ks}{}^i \Gamma_{jm}{}^s).$$

The condition for zero torsion then becomes

$$\Gamma_{ij}{}^k = \Gamma_{ji}{}^k.$$

For this reason connexions with torsion zero are referred to as symmetric connexions. The notation used is that given by Nomizu [66], except for a change in the sign of T and R.

Finally, we define the functions $\Gamma_{ij}{}^k$ and $R^i{}_{mjk}$ directly in terms of the connexion. Consider the coordinate cross section χ of $B(M)$ defined by

$$\chi(n) = (n, X_1(n), ..., X_d(n)).$$

Then

$$R^i{}_{mjk} = \Phi_{im}(d\chi\, X_j , d\chi\, X_k),$$

and

$$\Gamma_{ij}{}^k = \phi_{kj}(d\chi\, X_i),$$

where ϕ_{kj} and Φ_{im} are the components of these $\mathfrak{gl}(d, R)$-valued forms.

A knowledge of the $\Gamma_{ij}{}^k$ for each coordinate system of a family whose coordinate neighborhoods cover M determines the connexion, since ϕ is determined on the cross sections [compare the remark following (iii) in 6.4.2].

Problem 17. Let χ be the cross section as above and let ρ, ψ, P, Ψ be the pulled down forms $\omega \circ d\chi$, $\phi \circ d\chi$, $\Omega \circ d\chi$, and $\Phi \circ d\chi$. Obtain the following formulas for these forms in the coordinate neighborhood:

$$\rho = (dx_1, ..., dx_d)$$

$$\psi_{jk} = \sum_i \Gamma_{ik}{}^j \, dx_i$$

$$P_k = \sum_{i,j} \Gamma_{ij}{}^k \, dx_i \, dx_j$$

$$\Psi_{im} = \sum_{j,k} R^i{}_{mjk} \, dx_j \, dx_k.$$

Thus derive the equation given for $R^i{}_{mjk}$ in terms of the $\Gamma_{ij}{}^k$ by using the pulled-down structural equation

$$d\psi = -\tfrac{1}{2}[\psi, \psi] + \Psi.$$

Problem 18. Pull down the formula $d\Phi = -[\phi, \Phi]$ (corollary, 5.5) to get the coordinate form of the Bianchi identity

$$R^m{}_{nij,k} + R^m{}_{njk,i} + R^m{}_{nki,j} = 0.$$

Problem 19. Prove that if x_i are normal coordinates at m, then $\Gamma_{ij}{}^k(m) + \Gamma_{ji}{}^k(m) = 0$.

Problem 20. Prove that torsion is zero if and only if for every m there is a coordinate system at m such that $\Gamma_{ij}{}^k(m) = 0$.

Problem 21. *Connexions and action of groups.* Let G act on M to the left in such a way that if $g \in G$ and dg is the identity on some M_m, then g is the identity of G. Choose $b \in B(M)$, $b = (m, e_1, ..., e_d)$ and define $I_b : G \to B(M)$, $I_b(g) = (gm, dg\,e_1, ..., dg\,e_d)$.

Show that I_b is an imbedding.

Let G also act on $B(M)$ by $gb = I_b(g)$. Then the action is by bundle maps. An affine connexion on M is *invariant under* G if the connexion form ϕ is invariant, $\phi \circ dg = \phi$, for every $g \in G$.

If M is a homogeneous space of G, show that an invariant connexion on (G, M, H) (see problem 5.5) induces an invariant affine connexion on M. In this case, if we write $\mathfrak{g} = \mathfrak{m} + \mathfrak{h}$, as before, then the imbedding takes $\mathfrak{h}$ into fundamental vector fields, $\mathfrak{m}$ into all basic vector fields, restricted to $I_b(G)$.

Problem 22. *Product connexions.* Let M', M'' be manifolds with affine connexions having connexion forms ϕ', ϕ'', solder forms ω', ω'', etc. Let $M = M' \times M''$ and define the *bundle of adapted bases over M to be the submanifold of $B(M)$*:

$$B(M', M'') = \{((m', m''), e_1, ..., e_d) \mid (m', e_1, .., e_{d'}) \in B(M')$$

$$\text{and} \quad (m'', e_{d'+1}, ..., e_d) \in B(M'')\}.$$

The group of $B(M', M'')$ is $Gl(d') \times Gl(d'')$ and it is clear that $B(M', M'')$ may be identified with $B(M') \times B(M'')$. Define a connexion $\phi' \oplus \phi''$ on $B(M', M'')$ and extend by equivariance to a

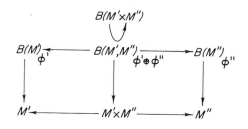

FIG. 23.

connexion ϕ on $B(M)$. M with this affine connexion is called the *affine product* of affinely connected manifolds M' and M''. Show that the product connexion has the following properties:

(a) If γ' and γ'' are curves in M' and M'', X' and X'' are parallel vector fields along γ' and γ'', then $X' + X''$ is parallel along $\gamma' \times \gamma''$, and, conversely, a parallel vector field on M has this form.

(b) The geodesics are products of geodesics on M' and M''. Hence the affine product of complete connexions is complete.

(c) $$T_{s'+s'',t'+t''} = T'_{s',t'} + T''_{s'',t''}.$$

(d) $$R_{s'+s'',t'+t''} = R'_{s',t'} + R''_{s'',t''}.$$

Problem 23. If $i : N \to M$ is a covering map, then there is an induced natural covering map $\bar{i} : B(N) \to B(M)$. (The *prolongation* of i.) If M has an affine connexion and if ϕ is the connexion form, then $\bar{i}^*\phi$ is a connexion form on $B(N)$. Describe this connexion in terms of parallel translation.

Problem 24. *Connexions on the affine bundle.* Let

$$A(d, R) = \left\{ \begin{pmatrix} A & x \\ 0 & 1 \end{pmatrix} \in Gl(d + 1, R) \mid A \in Gl(d, R),\ x \in R^d \right\}.$$

(x is viewed as a column matrix.)

$$A(M) = \{(b, t) \mid b \in B(M),\ t \in M_{\pi(b)}\}.$$

Define a right action $A(M) \times A(d, R) \to A(M)$ by

$$(b, t) \begin{pmatrix} A & x \\ 0 & 1 \end{pmatrix} = (bA, bx + t).$$

Show that this defines $A(M)$ as a principal bundle over M. (See proof of theorem 10.)

If we view R^d as the hyperplane of R^{d+1} which has final coordinate 1, then $A(d, R)$ acts to the left on R^d:

$$\begin{pmatrix} A & x \\ 0 & 1 \end{pmatrix} y = Ay + x, \quad \text{for} \quad y \in R^d.$$

Hence there is an associated bundle $S(M)$ which has fibres homeomorphic to the tangent spaces of M. Make this correspondence explicit.

Define maps

$$\eta : B(M) \to A(M)$$
$$\eta_G : Gl(d, R) \to A(d, R)$$

by $\eta(b) = (b, 0)$, $\eta_G(A) = \begin{pmatrix} A & 0 \\ 0 & 1 \end{pmatrix}$, and show that this gives a bundle map of $B(M)$ into $A(M)$ which induces the identity map on the base space M.

Let ω be the solder form on $B(M)$ and consider a connexion ϕ on $B(M)$, with Φ and Ω as the curvature and torsion forms.

The Lie algebra of $A(d, R)$ may be considered to be

$$\mathfrak{a}(d, R) = \left\{ \begin{pmatrix} X & x \\ 0 & 0 \end{pmatrix} \mid X \in \mathfrak{gl}(d, R),\ x \in R^d \right\}.$$

Define an $\mathfrak{a}(d, R)$-valued form $\bar{\phi}$ on $\eta(B(M))$ with the property that $\eta^*\bar{\phi} = \begin{pmatrix} \phi & \omega \\ 0 & 0 \end{pmatrix}$, and show that $\bar{\phi}$ may be extended by right translation to a connexion form on $A(M)$, also denoted by $\bar{\phi}$. If $\bar{\Phi}$ is the corresponding curvature form, show that

$$\eta^*\bar{\Phi} = \begin{pmatrix} \Phi & \Omega \\ 0 & 0 \end{pmatrix}.$$

A connexion on $A(M)$ arising in this way from a connexion on $B(M)$ is a special case of a Cartan connexion (see [35, 48, 49]). By considering other horizontal, R^d-valued equivariant forms in place of ω, more general Cartan connexions may be defined, and in general all the connexions on $A(M)$ whose distributions are disjoint from $T(\eta(B(M)))$ arise in this way, since $\mathfrak{gl}(d, R)$ is reductive in $\mathfrak{a}(d, R)$.

Returning to the connexion $\bar{\phi}$, we note that the parallel translation induced in the associated bundle $S(M)$ gives rise to affine transformations of the tangent spaces to M which depend on both the curvature and torsion of the connexion ϕ on $B(M)$.

Infinitesimally, a curvature transformation $\bar{R}$ may be defined in a way analogous to the curvature transformation of an affine connexion as follows.

Let x, y, $z \in M_m$, $(b, t) \in A(M)$ such that $\pi(b, t) = m$, $\bar{x}$, $\bar{y}$ lifts of x, y to (b, t). Then

$$\bar{R}_{xy} z = -(b, t) \, \bar{\Phi}(\bar{x}, \bar{y})(b, t)^{-1}z,$$

where we are identifying the fibre of $S(M)$ at m with M_m.

By choosing $(b, t) \in \eta(B(M))$, show that

$$\bar{R}_{xy} z = R_{xy} z + T_{xy}.$$

CHAPTER 7

Riemannian Manifolds

The definition of a Riemannian structure on a manifold is given and the corresponding topological metric is shown to induce the same topology. The bundle of (orthonormal) frames is defined and the existence and uniqueness of the Riemannian connexion is established. The chapter concludes with a large selection of examples [*33, 50, 83*].

7.1 Definitions and First Properties

7.1.1 Riemannian Metrics and Associated Topological Metrics

A *Riemannian manifold* is a manifold M for which is given at each $m \in M$ a positive definite symmetric bilinear form $\langle \, , \, \rangle$ on M_m, and this assignment is C^∞ in the sense that for any coordinate system $(x_1, ..., x_d)$ the functions $g_{ij} = \langle D_{x_i}, D_{x_j} \rangle \in C^\infty$. Such an assignment is called a *Riemannian metric* on M.

If we let $Sy(M)$ be the bundle of symmetric positive definite tensors of type $(2, 0)$, then a more elegant version of the above definition is that a Riemannian manifold is one with a preferred C^∞ cross section of this bundle $Sy(M)$.

Let M, N be Riemannian manifolds with metrics $\langle \, , \, \rangle_M$ and $\langle \, , \, \rangle_N$. Then a C^∞ map $f : M \to N$ is an *isometry* if it is a homeomorphism and preserves the metrics, that is, for $t, s \in M_m$, $\langle dft, dfs \rangle_N = \langle t, s \rangle_M$. An isometry is a diffeomorphism. f is called a *local isometry* if we relax the requirement that it be one-to-one.

If M is an oriented Riemannian manifold then there is a unique d-form θ which determines the orientation and such that $\theta(e_1, ..., e_d) = \pm 1$ for every orthonormal basis $e_1, ..., e_d$ of M_m. θ is called the *Riemannian volume element* of the oriented Riemannian manifold.

We defer examples in this chapter to the last section.

If $(x_1, \ldots, x_d)$ is a coordinate system on any manifold M with domain O, then there is a natural inner product on the tangent spaces to O, namely, the Euclidean inner product $\langle D_{x_i}, D_{x_j} \rangle = \delta_{ij}$. We denote by $\| \ \|'$ the Euclidean norm.

We also let $\| \ \|$ be the Riemannian norm, that is, $\| t \| = \langle t, t \rangle^{1/2}$, $t \in T(M)$, so that we have $\| D_{x_i} \| = (g_{ii})^{1/2} \in C^\infty$, and in fact for $a_i \in C^\infty$, $\| \sum_i a_i D_{x_i} \| \in C^\infty$ at each point at which not all the a_i vanish.

If we require $\langle \ , \ \rangle$ to be only nondegenerate instead of positive definite, then M is called a *semi-Riemannian manifold*. The main result of this chapter, namely, the existence and uniqueness of a Riemannian connexion, holds in the semi-Riemannian case.

Problem 1. The index of a symmetric quadratic form on a real vector space is the dimension of a maximal subspace on which the form is negative definite. Prove that for a connected semi-Riemannian manifold the index of the metric is the same on every tangent space.

For nonconnected manifolds we also require that the index be constant for a semi-Riemannian metric.

A manifold with index 1 or $d - 1$ is called a *Lorentz manifold*. The four-dimensional time-space universe of Einstein is a Lorentz manifold.

If γ is a broken C^∞ curve in M, then its *arc length* is defined by

$$| \gamma | = \int_a^b \| \gamma_* \|,$$

where $[a, b]$ is the interval of definition of γ.

Problem 2. Let γ be a broken C^∞ curve defined on $[a, b]$. Define nondecreasing continuous function f on $[a, b]$ by

$$f(x) = \int_a^x \| \gamma_* \|.$$

(a) Show that f is C^∞ at every x such that $\gamma_*(x)$ exists and is nonzero.

(b) Show that $\gamma \circ f^{-1} : [0, | \gamma |] \to M$ is a continuous well-defined function even though f^{-1} may not be a function, and that it is C^∞ at every $f(x)$ for which $\gamma_*(x) \neq 0$.

(c) Let (x, y) be a coordinate system on a two-dimensional manifold and define C^∞ curve γ on an interval with 0 as an interior point by the equations

$$x(\gamma(t)) = \int_0^t [\exp(-1/s^2)\sin 1/s]^2\, ds, \quad y(\gamma(t)) = \int_0^t \exp(-1/s^2)\sin 1/s\, ds.$$

Show that for this curve $\gamma \circ f^{-1}$ is not a broken C^∞ curve so that γ cannot be reparametrized with respect to arc length so as to remain broken C^∞.

We define a function $\rho : M \times M \to R \cup \{+\infty\}$ by

$$\rho(m, n) = \inf_{\gamma \in \Gamma} |\gamma|,$$

where $\Gamma =$ set of all broken C^∞ curves from m to n.

Lemma 1.   The function ρ is a metric on M.

Proof.   It is trivial that ρ is symmetric and satisfies the triangle inequality, so the only thing remaining to be proved is that $\rho(m, n) = 0$ implies that $m = n$. Assume that $m \neq n$, and let $(x_1, ..., x_d)$ be a coordinate system at m with domain U. Let O be a ball with respect to the x_i such that $\bar{O} \subset U$ and $n \notin O$. Define a function $f : R^d \times U \to R$ by

$$f(a_1, ..., a_d, m) = \left\| \sum a_i D_{x_i}(m) \right\|.$$

Then $f\,|_{S^{d-1} \times U}$ is continuous and positive, and therefore there exists a $k > 0$ such that

$$\frac{1}{k} \leqslant f\,|_{S^{d-1} \times \bar{O}} \leqslant k.$$

Now $\left\| \sum a_i D_{x_i}(m) \right\|' = 1$ on $S^{d-1} \times O$, and hence we have

$$\frac{1}{k}\left\| \sum a_i D_{x_i}(m) \right\|' \leqslant \left\| \sum a_i D_{x_i}(m) \right\| \leqslant k \left\| \sum a_i D_{x_i}(m) \right\|' \tag{1}$$

for $(a_1, ..., a_d, m) \in S^{d-1} \times \bar{O}$. Then by the fact that the expressions in (1) are homogeneous in the a_i's, we have that (1) holds for $(a_1, ..., a_d, m) \in R^d \times \bar{O}$. Now let γ be any broken C^∞ curve from

FIG. 24.

m to n, and let γ' be the part of γ from m to the first place where γ intersects the boundary of O. Finally, let α be the radius of O. We then have

$$\rho(m, n) = \inf |\gamma| \geqslant \inf |\gamma'| \geqslant \frac{1}{k} \inf |\gamma'|' \geqslant \frac{1}{k} \alpha > 0,$$

which completes the proof of the lemma.

Problem 3. In the proof of the lemma we have assumed the theorem about Euclidean space that says $|\sigma|' \geqslant \alpha$ for every curve σ which goes from the origin to a point on the sphere of radius α. Prove this result, and also that equality occurs only if σ is a broken C^∞ reparametrization of a straight line. [*Hint*: Let r be the function $(\sum x_i^2)^{1/2}$. Split $\sigma_*(t)$ into two components, one, σ_{*T} tangent to the ray from the origin, the other normal to that ray. Show that $|d(r \circ \sigma)/dt| = \| \sigma_{*T} \|$, and apply the fundamental theorem of calculus. Compare the similar theorem for geodesics in a Riemannian manifold in Section 8.1.]

Theorem 1. The topology given by ρ is equivalent to the topology of M as a manifold. Hence, ρ is a continuous function on $M \times M$.

Proof. It is sufficient to find for each $m_0 \in M$ a neighborhood P whose topology is given by ρ. Let $x_1, \ldots, x_d$ be a coordinate system at m_0, O be an open ball with respect to the x_i with center at m_0, O_1 similarly, with $\bar{O}_1 \subset O$. Let ρ' be the Euclidean metric on O defined via the x_i, so that ρ' defines the topology on O. Hence, we wish to find a neighborhood P of m_0 and a number $c > 0$ such that $P \subset O$ and $(1/c) \rho' \leqslant \rho \leqslant c\rho'$ on $P \times P$. From Eq. (1) above, we have that there exists $c > 0$ such that

$$\frac{1}{c} \| t \|' \leqslant \| t \| \leqslant c \| t \|'$$

for $t \in M_m$, $m \in \bar{O}_1$. Hence for any broken C^∞ curve in O_1, we have $(1/c) |\gamma|' \leqslant |\gamma| \leqslant c |\gamma|'$. So we need only worry about curves which leave O_1. This we do by cutting down again. Let $\delta_1 = \rho'$-radius of O_1, and take $\beta \leqslant \delta_1/(2c^2 + 1)$. Let P be the open ball of radius β with respect to the x_i about m_0. Now from the above remark, it is clear that in order to show

$$\frac{1}{c} \rho' \leqslant \rho \leqslant c\rho'$$

on $P \times P$, we need only prove that in the calculation of ρ on $P \times P$ we can restrict ourselves to curves in O_1. That is, if $(m, n) \in P \times P$ and γ is a curve from m to n, then there exists a curve $\bar{\gamma}$ such that $|\bar{\gamma}| \leqslant |\gamma|$ and $\bar{\gamma}$ is in O_1. Now let $\gamma_1 =$ part of γ from m to the first intersection point of γ with the boundary of O_1, and let $\bar{\gamma} =$ the straight line from m to n with respect to the x_i. We then have

$$|\gamma| \geqslant |\gamma_1| \geqslant \frac{1}{c}|\gamma_1|' \geqslant \frac{1}{c}(\delta_1 - \beta) \geqslant 2c\beta \geqslant c|\bar{\gamma}|' \geqslant |\bar{\gamma}|. \quad \text{QED}$$

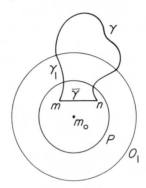

FIG. 25.

Before discussing further properties of Riemannian manifolds, we shall show that a large class of manifolds admits Riemannian structures.

Theorem 2. If M is a paracompact manifold, then M admits a Riemannian metric.

Proof. Let $\{U_i\}$ be a covering of M by coordinate systems, and let $\{f_i\}$ be a C^∞ locally finite partition of unity subordinate to this covering. Let $(x_1{}^i, ..., x_d{}^i)$ be the coordinate system associated with U_i, and denote by $\langle\,,\,\rangle_i$ the Euclidean inner product defined on U_i via the $x_j{}^i$. We then define an inner product on M by $\langle\,,\,\rangle = \Sigma f_i \langle\,,\,\rangle_i$.

Remark. There is a similar result for real-analytic manifolds arising from the solution of the analytic imbedding problem [31, 56].

7.1.2 Vector Bundles

Let (B, R^n, M) be an n-dimensional vector bundle over M (3.3(4)), with B the bundle space and R^n the fibre. A *Riemannian metric* on the vector bundle B is a C^∞ assignment of a symmetric, positive definite bilinear form to each fibre. By C^∞ we mean the following. Let $U \subset M$, χ_1 and χ_2 be C^∞ cross sections of U in B. Then the function $f : U \to R$ given by $f(m) = \langle \chi_1(m), \chi_2(m) \rangle$ is C^∞.

By a proof similar to the above it can be shown that an n-dimensional vector bundle B over a paracompact base M admits a Riemannian metric. In fact, let $U \subset M$ be an open set such that $\pi^{-1}(U)$ is diffeomorphic to $R^n \times U$. Then by considering a basis of R^n it is easy to see that there exist n linearly independent C^∞ cross sections $\chi_1, ..., \chi_n$ of U into $\pi^{-1}(U)$. We therefore can define a Riemannian metric on $\pi^{-1}(U)$ by setting $\langle \chi_i, \chi_j \rangle = \delta_{ij}$. Now using a covering of M by open sets over which B is trivial and an associated partition of unity, the proof goes through in the same way.

7.2 The Bundle of Frames

Let M be a Riemannian manifold. We have discussed the bundle of bases $B(M)$ over M. Let

$$F(M) = \{(m, e_1, ..., e_d) \mid m \in M, e_1, ..., e_d \text{ an orthonormal basis of } M_m\},$$

and let $\pi' : F(M) \to M$ be the obvious projection. $F(M) \subset B(M)$. We shall put a local product structure on $F(M)$ so that it becomes a manifold, a submanifold of $B(M)$, and a principal bundle over M, called the *bundle of frames*, which represents a reduction of the group of $B(M)$ to the orthogonal group.

Let $m \in M$, $(x_1, ..., x_d)$ be a coordinate system at m with domain U. We define a function

$$\lambda_U : U \to Gl(d, R).$$

Let $V_1, ..., V_d$ be vector fields defined on U with the property that for all $n \in U$, $V_1(n), ..., V_d(n)$ is the Gram-Schmidt orthonormalization of $D_x(n), ..., D_{x_d}(n)$. Then $\lambda_U(n)$ is defined by

$$V_i(n) = \sum_{j=1}^{i} (\lambda_U(n))_{ij} D_{x_j}(n).$$

Now we define the following maps,

$$\phi_U : \pi^{-1}(U) \to Gl(d, R),$$

$$\phi_U' : \pi'^{-1}(U) \to O(d),$$

as follows:

$$e_i = \sum (\phi_U(n, e_1, ..., e_d))_{ij} V_j(n)$$

$$f_i = \sum (\phi_{U'}(n, f_1, ..., f_d))_{ij} V_j(n)$$

$i = 1, ..., d$, where $(n, e_1, ..., e_d) \in B(M)$, $(n, f_1, ..., f_d) \in F(M)$. Proceeding similarly for a covering of M by coordinate neighborhoods and defining right action exactly as in 3.2, we see that we have defined principal bundle structures on $B(M)$ and $F(M)$, and the local product representations are compatible with the differential structure. In particular, $F(M)$ is a manifold and has structural group $O(d)$, while the λ_U, being clearly C^∞, show that the bundle structure defined on $B(M)$ is the same as that introduced previously. Further, we have immediately from 3.4 that $F(M)$ represents a reduction of the group $Gl(d, R)$ of $B(M)$ to $O(d)$, and $F(M)$ is a submanifold of $B(M)$.

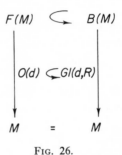

FIG. 26.

Since $\mathfrak{o}(d)$ may be viewed as the set of all $d \times d$ skew-symmetric real matrices, it clearly admits $\{X_{ij} - X_{ji} \mid i < j\}$ as a basis, where X_{ij} is defined as in 6.1.2. Hence, $\lambda\mathfrak{o}(d) = \bar{\mathfrak{o}}(d)$ admits as a basis $\{F_{ij} = E_{ij} - E_{ji} \mid i < j\}$. In other words, at each $f \in F(M)$, the vertical tangent space to $F(M)$, namely, $V_f' = V_f \cap F(M)_f$, where V_f is the vertical tangent space to $B(M)$, is spanned by the tangents

$$\{F_{ij}(f) = E_{ij}(f) - E_{ji}(f) \mid i < j\}.$$

If H' is a connexion on $F(M)$, then it can clearly be extended to a connexion H on $B(M)$ by right action. Hence, H gives rise to a parallel translation of tangents to M along curves in M. Further, it is clear that this parallel translation preserves scalar products, for it is defined via $F(M)$ by sending one orthonormal base into another.

Conversely, if H is a connexion on $B(M)$ such that parallel translation preserves scalar products, then H comes from a connexion H' on $F(M)$ in the above manner. Indeed, let $b = (m, f_1, ..., f_d) \in F(M)$ and let γ be a horizontal curve in $B(M)$ passing through $(m, f_1, ..., f_d)$. Then every point on γ must belong to $F(M)$, since parallel translation along $\pi \circ \gamma$ takes $f_1, ..., f_d$ into orthonormal bases, by assumption. Therefore $H_b \subset F(M)_b$, so we may define H' by $H'_b = H_b$.

Problem 4. Prove that a reduction of the group of $B(M)$ to $O(d)$ gives a unique Riemannian metric such that the reduced bundle is $F(M)$.

Problem 5. Extend the results of this section to apply to a Riemannian metric on an arbitrary vector bundle.

Remark. In a similar way, a semi-Riemannian structure on M gives a reduction of $B(M)$ to a subgroup of $Gl(d, R)$ which leaves invariant a non-degenerate symmetric bilinear form, and conversely.

7.3 Riemannian Connexions

Let M be a Riemannian manifold. A connexion on $B(M)$ is called a *Riemannian connexion* if it satisfies the following properties:

(i) parallel translation preserves inner products,
(ii) the torsion form is zero.

We note that in view of the above remarks, a connexion on $B(M)$ is a Riemannian connexion if and only if it is the extension of a connexion of $F(M)$ whose torsion form is zero. We defined the solder form ω on $B(M)$ in the last chapter. If $i : F(M) \subseteq B(M)$ is the inclusion map, then we again denote by ω the horizontal 1-form $i^*\omega$ defined on $F(M)$. If H' is a connexion on $F(M)$, H its extension to $B(M)$, and if we denote the corresponding 1-forms by ϕ and ϕ', respectively, then the first structural equation (theorem 6.4) pulls back to $F(M)$; namely,

$$d\omega = -\phi'\omega + i^*\Omega.$$

Hence, we have that H is a Riemannian connexion if and only if it comes from a connexion on $F(M)$, also called a Riemannian connexion, whose 1-form ϕ' satisfies the relation $d\omega = -\phi'\omega$.

Lemma 2. There is a one-to-one correspondence between horizontal $\mathfrak{o}(d)$-valued 1-forms τ on $F(M)$ and horizontal R^d-valued 2-forms, in such a way that the 2-form associated with τ is $\tau\omega$. When τ is not horizontal it is still determined by $\tau\omega$.

Proof. We first show that for a horizontal 2-form Ω there is a unique horizontal 1-form τ such that $\Omega = \tau\omega$. Let $K(x, y, z) = \langle \Omega(x, y), \omega(z)\rangle$, so that K is 3-linear, horizontal, and skew-symmetric in the first two variables. Since ω is onto R^d, there is a unique $\tau(x) \in \mathfrak{gl}(d, R)$, $x \in F(M)_f$, such that for every y, $z \in F(M)_f$

$$2 \langle \tau(x) \omega(y), \omega(z)\rangle = K(x, y, z) + K(z, y, x) - K(x, z, y).$$

This does not overdetermine $\tau(x)$, because if we alter y and z by vertical vectors, so that $\omega(y)$ and $\omega(z)$ do not change, then the right-hand side will not change either. τ is obviously horizontal.

Since interchanging y and z on the right-hand side changes its sign, we must have $\tau(x) \in \mathfrak{o}(d)$. To show that $\tau\omega = \Omega$, it suffices to show that

$$K(x, y, z) = \langle \tau(x) \omega(y), \omega(z)\rangle - \langle \tau(y) \omega(x), \omega(z)\rangle.$$

This is true because the second plus the third term of the right-hand side above is invariant under interchange of x and y, while the first term changes sign.

When τ is not horizontal then using $\tau\omega$ in place of Ω in defining K, it is an automatic verification that $2\langle \tau(x) \omega(y), \omega(z)\rangle$ is given as above in terms of K, so $\tau\omega$ determines τ. QED

Remark. Lemma 2 may be proved by showing the map $\tau \to \tau\omega$ is one-to-one and then applying a dimensionality argument. The lemma also holds if the forms τ are taking values in a subalgebra consisting of all linear transformations which are skew-symmetric with respect to some non-degenerate symmetric bilinear form in R^d.

Theorem 3. There is a one-to-one correspondence between connexions on $F(M)$ and horizontal equivariant 2-forms on $F(M)$. The 2-form may be taken to be the torsion form of the connexion, so two connexions are equal if and only if they have the same torsion.

Proof. Let ϕ be a connexion form on $F(M)$. Fixing ϕ we get a one-to-one correspondence between connexion forms ψ on $F(M)$ and the difference forms $\tau = \psi - \phi$, which is an arbitrary $\mathfrak{o}(d)$-valued horizontal equivariant 1-form. The lemma shows that τ is determined by $\tau\omega$, and furthermore it is trivial to verify that τ is equivariant if and only if $\tau\omega$ is equivariant. This establishes a one-to-one correspondence between connexions on $F(M)$ and horizontal equivariant 2-forms $\tau\omega$. Finally, by the first structural equation we have

$$d\omega = -\phi\omega + \Omega, \qquad d\omega = -\psi\omega + \Omega_1 ,$$

so

$$\tau\omega = (\psi - \phi)\,\omega = \Omega_1 - \Omega,$$

that is, $\Omega_1 = \tau\omega + \Omega$. Thus by choosing $\tau\omega = -\Omega$ we get a connexion ϕ_0 with torsion 0. If we started with ϕ_0 instead of ϕ we would have $\Omega_1 = \tau\omega$, that is, the correspondence would then be between connexions and torsion forms. QED

Problem 6. Find a formula for ϕ_0 in terms of ω and $d\omega$ using the above lemma and the fact that $\phi_0\omega = -d\omega$.

Theorem 4. There exists a unique connexion on $F(M)$ having torsion 0; thus there exists a unique Riemannian connexion on $B(M)$. (This follows immediately from the previous proof.)

Remark. By the previous remark it is clear that by the same method we may prove the existence and uniqueness of Riemannian connexions (same definition) for semi-Riemannian structures.

Proposition. The connexions on $F(M)$ having the same geodesics as a given connexion are in one-to-one correspondence with horizontal equivariant 3-forms on $F(M)$.

Proof. The condition on the difference form τ in order that two connexions have the same geodesics is $\tau(x)\,\omega(y) = -\tau(y)\,\omega(x)$ (theorem 6.9). The 3-linear function $\langle\tau(x)\,\omega(y), \omega(z)\rangle$ is a 3-form if and only if τ satisfies this condition and is $\mathfrak{o}(d)$-valued. That it is horizontal and equivariant is easily checked. [Equivariance here means invariance under right action of $O(d)$.]

Problem 7. On a two-dimensional Riemannian manifold two different connexions on $F(M)$ have different geodesics.

Problem 8. Prove that an isometry of a connected Riemannian manifold is determined by its value and differential at one point as follows:

(a) First reduce to the problem of showing that an isometry which leaves a point and the tangent space at that point fixed is the identity.

(b) An isometry takes geodesics into geodesics.

(c) Show that an isometry which leaves a point and tangent space at the point fixed leaves a neighborhood of the point fixed.

(d) The collection of all fixed points of a continuous map is a closed set.

7.4 Examples and Problems

1. *Euclidean space.* For every $m \in R^d$ we define $\langle D_i(m), D_j(m) \rangle = \delta_{ij}$, thus defining the *flat Riemannian metric on* R^d. Unless otherwise stated, a metric on R^d will be assumed to be the flat one.

Problem 9. Show that the connexion on R^d given by its group structure as in problem 6.6 is the Riemannian connexion, so the Riemannian connexion is indeed flat. What are the geodesics?

2. *One-dimensional Riemannian manifolds.* Since Riemannian manifolds are metrizable they must be paracompact, so the underlying manifold of a connected one-dimensional Riemannian manifold must be either R or S^1. In either case there are only two vector fields having norm constantly 1, and an integral curve of such a vector field provides an isometry with an interval of R (as in example 1) or a covering map of S^1 by R which is locally an isometry and periodic as a function on R. In the latter case the smallest period is the *circumference* of S^1, divided by 2π it is the *radius* of S^1, since S^1 may be imbedded in R^2 as a circle of that radius (see example 3).

3. *Imbeddings.* If $i : N \to M$ is a C^∞ imbedding (or, more generally, an immersion) and M is a Riemannian manifold, then there is a unique Riemannian metric on N such that di preserves inner products; when N has this metric i is said to be an *isometric imbedding* (*immersion*). A very difficult theorem of Nash [63] says that every Riemannian manifold has an isometric imbedding in an arbitrarily small neighborhood of Euclidean space of dimension $d(d + 1)(3d + 11)/2$, and if it is compact, dimension $d(3d + 11)/2$. One of the harder outstanding problems is to what extent this theorem can be improved in the allowable gap in dimensions and in the amount of uniqueness possible. (Immersions are defined and discussed in Chapter 10.)

Problem 10. Take a circle of radius r and center $(s, 0, 0)$, $r < s$, and lying in the plane $u_2 = 0$ and rotate it about the u_3-axis. Realize this surface as the range of an imbedding of T^2, the 2-torus. Compute the Riemannian metric.

4. *The Riemannian d-sphere* of radius r, $S^d = \{x \in R^{d+1} \mid \| x \| = r\}$, is obtained by giving it the induced Riemannian metric from R^{d+1}. Rotations are all isometries, so S^d has $O(d + 1)$ as a transitive group of isometries.

Problem 11. Let $x_0, ..., x_{d-1}$ be linearly independent points of R^{d+1} all of which lie on S^d. Show that there exists a unique isometric imbedding of S^k, the k-sphere with radius r, in S^d whose range contains $x_0, ..., x_k$, $k = 1, ..., d - 1$. This gives a chain $S^1 \subset S^2 \subset \cdots \subset S^{d-1} \subset S^d$. Prove that any such chain can be mapped on any other by an element of $SO(d + 1)$.

Problem 12. Prove the following facts about S^d:

(a) For any $m \in S^d$, there is a unique isometry j_m leaving m fixed and such that $dj_m = -$ identity on S^d_m .

(b) For each nonzero $t \in S^d_m$, there is a unique great circle $S^1(t)$ tangent to t, and $S^1(t)$ is invariant under j_m , that is, $S^1(t) = j_m(S^1(t)) = S^1(-t)$.

(c) If X is a parallel vector field along $S^1(t)$, then $dj_m X = -X$.

(d) If m, n, p are equally spaced along $S^1(t)$ and $X(m) = t$, use j_n to show that $X(p) = $ a tangent to $S^1(t)$ at p, and hence $S^1(t)$ is a geodesic.

5. *Riemannian products.* If M' and M'' are Riemannian manifolds, then $M = M' \times M''$ becomes a Riemannian manifold when the inner products are defined as

$$\langle s' + s'', t' + t'' \rangle = \langle s', t' \rangle' + \langle s'', t'' \rangle'',$$

that is, tangents to one manifold are regarded as being perpendicular to those of the other manifold. M is then called the *Riemannian product of M' and M''*. For example, R^d is the d-fold Riemannian product of R.

Problem 13. Show that the Riemannian connexion of the Riemannian product of Riemannian manifolds is the affine product of the Riemannian connexions of the manifolds. (See problem 6.22.)

6. *Flat tori.* The Riemannian product of S^1 with itself d times is called the *flat d-dimensional torus T^d*. No distinction is usually made

between S^1 of different radii, and in a given torus all of the radii need not be the same.

Problem 14.   Show that the flat d-dimensional torus T^d is flat.

7. *Covering manifolds.*   If $i : N \to M$ is a covering map, M a Riemannian manifold, then i may be considered as an immersion, and when N is given the induced Riemannian metric, (i, N) is called a *Riemannian covering of M.* For example, the simply connected Riemannian covering space of a flat torus is R^d, since this is true for $d = 1$, and the relation is preserved under products.

Problem 15.   Prove the last remark, namely, that if $i' : N' \to M'$, $i'' : N'' \to M''$ are Riemannian coverings, then the Riemannian product $N' \times N''$ is the Riemannian covering of $M' \times M''$ with covering map $i' \times i''$.

Problem 16.   There is a reverse process, also. If $i : N \to M$ is a covering map and if N has a Riemannian structure for which the deck transformations are isometries, then there is a natural induced Riemannian structure on M which makes (i, N) into a Riemannian covering of M. For example, real projective space P^d is covered by S^d and can be given a structure from that on the S^d of example 4.

8. *Parallelizable manifolds.*   If $X_1, ..., X_d$ are parallelizing vector fields on a manifold M, then $\langle X_i, X_j \rangle = \delta_{ij}$ (constant functions) defines a corresponding Riemannian metric on M. R^d and T^d are special cases; more generally, any Lie group has such a metric.

Problem 17.   Show that the Riemannian connexion of this metric is the same as the direct or opposite connexions if and only if $[X_i, X_j] = 0$ for every i, j.

Problem 18.   The Riemannian connexion of a parallelization is the torsion free connexion (problem 6.5) if and only if for all constant linear combinations X, Y of the parallelizing fields $\langle [X, Y], X \rangle = 0$, which is equivalent to the condition

$$\langle [X_i, X_j], X_k \rangle + \langle X_j, [X_i, X_k] \rangle = 0$$

for every i, j, k. Prove this by piecing together the following:

(a) The Riemannian and torsion free connexions are the same if and only if the geodesics of the Riemannian connexion are the same as those of the direct connexion (corollary to theorem 6.9).

(b) Let θ be the $O(d)$-valued form on M obtained by pulling down the Riemannian connexion form via the natural cross section of the parallelization, and ρ the parallelizing form. Show that the two connexions are the same if and only if $\theta(t) \rho(t) = 0$ for every $t \in T(M)$. (Use theorem 6.9, noting that θ is a difference form pulled down to M.)

(c) Use the structural equation $d\rho = -\theta\rho$ and the skew-symmetry of θ to show the desired result. (Note $\langle \rho([X, Y]), \rho(X) \rangle = \langle [X, Y], X \rangle$.)

Notice that an affine connexion gives a parallelization of $B(M)$ and hence a Riemannian structure, and the same holds for a Riemannian connexion on $F(M)$. Show that in the latter case, the horizontal geodesics in $F(M)$ are precisely the lifts of geodesics in M, while the vertical geodesics are integral curves of the fundamental vector fields F_{ij}. However, integral curves of constant linear combinations of fundamental and basic vector fields are not in general geodesics.

9. *Homometries.* If $f : N \to M$ is a diffeomorphism of Riemannian manifolds, then f is called a *homometry* (with expansion factor a) if for every $s, t \in N_n$, any n, $\langle df(s), df(t) \rangle = a^2 \langle s, t \rangle$. If f is the identity map on a manifold, then for a given Riemannian metric on M (or N) this formula defines another metric so that f is a homometry.

We have already noted that S^1 may have different radii, and this amounts to changes of S^1 by homometries.

If $N = M$ as a Riemannian manifold, then there may be no homometries which are not isometries ($a = 1$). Indeed, this is the case when M is compact (more generally, has finite volume), for then a homometry would have to multiply volume by the factor a^d. (We have not defined volume, but it can be done along fairly usual measure theoretic lines.)

If M has infinite volume it may well have self-homometries. For example, on R^d multiplication by a is a homometry.

Problem 19. Show that a homometry $f : M \to N$, when extended to a map $\bar{f} : B(M) \to B(N)$ ($\bar{f}(m, e_1, ..., e_d) = (f(m), df(e_1), ..., df(e_d))$), called the *prolongation* of f to $B(M)$, preserves the Riemannian connexions of the homometrically related metrics on M and N. In particular, multiplication of a metric by a positive scalar does not alter the Riemannian connexion. However, two metrics with the same connexions need not differ by a scalar, as can be seen by considering a Riemannian product.

10. *Conformal maps.* If we allow a to be a positive C^∞ function of n in example 9, we get a *conformal map*

$$f : N \to M, \qquad \langle df(s), df(t) \rangle = (a(n))^2 \langle s, t \rangle, \qquad s, t \in N_n.$$

The linear fractional transformations of the Riemann sphere of complex analysis are conformal maps of the Riemannian 2-sphere, and the homometries of R^d combined with stereographic projection give conformal maps of the Riemannian d-sphere.

Problem 20. Prove that stereographic projection is a conformal map of $S^d - \{pt\}$ onto R^d.

11. *Action by compact groups.* If G is a compact Lie group then G has a unique (Haar) measure which is invariant under left and right translations and assigns 1 to all of G. If G acts differentiably on M, then we can obtain a metric on M so that G acts as a group of isometries. For let $\langle \, , \rangle$ be any Riemannian metric on M, and for $s, t \in M_m$, define

$$\langle s, t \rangle' = \int_{g \in G} \langle dg(s), dg(t) \rangle.$$

Problem 21. Show that G acts as isometries on $(M, \langle \, , \rangle')$. If f is any self-isometry of $(M, \langle \, , \rangle)$ which commutes with the action of G, that is, $fg = gf$ for every $g \in G$, show that f is an isometry of $(M, \langle \, , \rangle')$.

Problem 22. Show that the orthogonal complements of the vertical tangents with respect to a right invariant metric on a principal bundle form a connexion distribution. Hence, give an alternate proof that a paracompact principal bundle admits a connexion (compare 5.4).

Problem 23. (a) Show that a left invariant metric on a Lie group G corresponds to an inner product on the Lie algebra $\mathfrak{g}$.

(b) A left invariant metric on G is also right invariant (and so adjoint invariant) if and only if the corresponding inner product $\langle \, , \rangle$ on $\mathfrak{g}$ is *invariant*, that is, if $X, Y, Z \in \mathfrak{g}$, then

$$\langle [X, Y], Z \rangle + \langle Y, [X, Z] \rangle = 0,$$

that is, $ad\, X$ is skew-symmetric. (This is simply a restatement of the fact that the function $\langle Ad\, e^{tX}Y, Ad\, e^{tX}Z \rangle$ is constant if and only if its derivative is zero. Compare with problem 2.5.)

(c) If G is compact, it always admits such a metric.

(d) The Riemannian connexions of such invariant metrics on G are all equal to the torsion free connexion of problem 6.6, and hence are complete (see problem 18).

(e) The *Killing form* of a Lie group G is a bilinear form $k(,)$ on $\mathfrak{g}$ defined by: if $X, Y \in \mathfrak{g}$, then

$$k(X, Y) = \text{tr}(\text{ad } X \circ \text{ad } Y).$$

$k(,)$ satisfies the invariance property of (b), but is not in general definite or even nondegenerate, for example, if g has a nontrivial center.

12. *Riemannian homogeneous spaces.* If M is acted upon by a transitive Lie group of isometries, then M is a *Riemannian homogeneous space.* By example 11, a homogeneous space of a compact Lie group may be given a homogeneous metric.

Problem 24. Let H be a closed reductive subgroup of a Lie group G (see problem 5.5), so that $\mathfrak{g} = \mathfrak{h} + \mathfrak{m}$, where $[\mathfrak{h}, \mathfrak{m}] \subset \mathfrak{m}$. Assume $\mathfrak{m}$ admits an inner product which is invariant under $Ad\, H$. Then show that G/H is a Riemannian homogeneous space.

$V_{d,r}$, the *Stiefel manifold* of ordered sets of r orthonormal vectors in R^d, is a Riemannian homogeneous space of both $O(d)$ and $SO(d)$, namely,

$$V_{d,r} = O(d)/O'(d - r) = SO(d)/SO'(d - r),$$

where $O'(d - r)$ and $SO'(d - r)$ are viewed as acting on the last $d - r$ components in R^d. If we make the definition $O(0) = SO(0) = \{1\}$, then $O(d) = V_{d,d}$, $SO(d) = V_{d,d-1}$, and $S^{d-1} = V_{d,1}$ are special cases.

13. *Flag manifolds.* If $d_1, ..., d_n$ is a partition of d, then we define flag manifold $Fl(d ; d_1, ..., d_n)$ as the set of n-tuples $(V_1, ..., V_n)$, where V_i is a subspace of R^d of dimension d_i and these subspaces are mutually orthogonal. Alternatively, $Fl(d ; d_1, ..., d_n)$ may be considered to be the set of increasing sequences

$$\{0\} = W_0 \subset W_1 \subset W_2 \subset \cdots \subset W_{n-1} \subset W_n = R^d$$

of subspaces of R^d with $d_i = \dim W_i - \dim W_{i-1}$.

Problem 25. (a) Establish a natural one-to-one correspondence between these two sets.

(b) The orthogonal group $O(d)$ acts on n-tuples $(V_1, ..., V_n)$ and the general linear goup $Gl(d, R)$ acts on increasing sequences

$W_1 \subset W_2 \subset \cdots \subset W_{n-1} \subset W_n = R^d$. Find the isotropy group in each case, thus giving $Fl(d \; ; d_1 , ..., d_n)$ the structure of a homogeneous space in two ways.

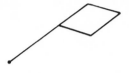

FIG. 27.

(c) Since $O(d)$ is compact, there is a Riemannian metric on $Fl(d \; ; d_1 , ..., d_n)$ on which $O(d)$ acts transitively as isometries. For $p = (V_1 , ..., V_n) \in Fl(d \; ; d_1 , ..., d_n)$ and $1 \leqslant i < j \leqslant n$, define m_{ij} to be the subspace of $Fl(d \; ; d_1 , ..., d_n)_p$ spanned by tangents to curves on which only V_i and V_j are varied. Show that these m_{ij} must be mutually orthogonal by using the invariance of the adjoint action of the isotropy algebra.

14. *Riemannian symmetric spaces* [*13, 18, 33*]. If a Riemannian manifold M has an isometry f_m for every $m \in M$ which leaves m fixed and such that $df_m |_{M_m} = -$ identity (f_m is called the *symmetry at m*) then M is a *Riemannian symmetric space*.

Since an isometry must take geodesics into geodesics, it is easy to see by a step-by-step use of symmetries at points along a geodesic that geodesics are infinitely extendable, so M is complete. If M is connected, then any two points of M may be joined by a broken geodesic (see problem 8), and hence the composition of the symmetries about the midpoints of the geodesic segments is an isometry sending one of the points into the other. Thus the group of isometries is transitive; and since this group is always a Lie group, M is a Riemannian homogeneous space [*50*].

S^d is a symmetric space.

Problem 26.   Prove that for a Riemannian symmetric space any C^∞ curve invariant under the symmetry about each of its points is a reparametrization of a geodesic. (Compare problem 12.)

Problem 27.   A Riemannian homogeneous space with a symmetry f_m at one point is Riemannian symmetric.

Problem 28.   *Alternate definition of Riemannian symmetric spaces.* Let M be a Riemannian homogeneous space, $M = G/H$. Then M is a *homogeneous Riemannian symmetric space* if G admits an automorphism f with the properties

(i) $f^2 = $ identity,

(ii) H is pointwise fixed under f and contains the greatest connected subgroup pointwise fixed under f.

(a) Let $\pi : G \to M$ be the canonical projection, and define $f_0 : M \to M$ by: $f_0(\pi(g)) = \pi(f(g))$. Show that f_0 is well-defined and $df_0 = -$ identity on M_0, where $0 = \pi(e)$.

(b) Use the formula $f_0 \circ L_g = L_{f(g)} \circ f_0$, where L denotes the left-action of G on M, and the fact that L_g is an isometry to show that f_0 is an isometry. Hence, prove that every homogeneous Riemannian symmetric space is a Riemannian symmetric space.

(c) Let M be a connected Riemannian symmetric space. We have already seen that M is Riemannian homogeneous, $M = G/H$, where $H = \{g \in G \mid g(m) = m\}$, m fixed $\in M$. Define $f : G \to G$ by: $f(g) = f_m g f_m^{-1}$. Show that f satisfies properties (i) and (ii) of a homogeneous Riemannian symmetric space. (See problem 8.)

Problem 29. Let G be a Lie group with a two-sided invariant metric. Show that the inverse map $\psi : G \to G$, defined by $\psi(g) = g^{-1}$, is an isometry, and hence prove that G is a Riemannian symmetric space.

G may also be exhibited as a homogeneous Riemannian space of the group $G \times G$, with the automorphism f defined by $f(g, h) = (h, g)$. Find the isotropy group and make the identification of G with the homogeneous space. In particular, describe the action of $G \times G$ on G.

Problem 30. *Grassmann manifolds.* $G_{d,r}$, the Grassmann manifold of r-planes in R^d, is a Riemannian homogeneous space of $O(d)$, namely,

$$G_{d,r} = \frac{O(d)}{O(r) \times O'(d - r)},$$

where $O(r)$ is viewed as acting on the first r components in R^d and $O'(d - r)$ on the last $d - r$. Let $g \in O(d)$, so that

$$g = \begin{pmatrix} A & B \\ C & D \end{pmatrix},$$

where A is $r \times r$, D is $(d - r) \times (d - r)$, and define f by

$$f(g) = \begin{pmatrix} A & -B \\ -C & D \end{pmatrix}.$$

Prove that f defines $G_{d,r}$ as a homogeneous Riemannian symmetric space.

$$\tilde{G}_{d,r} = \frac{SO(d)}{SO(r) \times SO'(d-r)},$$

the Grassmann manifold of oriented r-planes in R^d, is a twofold covering of $G_{d,r}$ and is also symmetric.

Problem 31. Let M and N be Riemannian symmetric spaces, $i : P \to M$ a covering map. Show that $M \times N$ and P with the induced Riemannian structures are Riemannian symmetric spaces.

Problem 32. *Lens spaces.* Consider the map

$$(z_1, z_2) \to (z_1 \exp 2\pi i/p,\ z_2 \exp 2\pi i q/p)$$

on

$$S^3 = \{(z_1, z_2) \mid z_i \in C,\ |z_1|^2 + |z_2|^2 = 1\},$$

where p and q are relatively prime integers. This map generates a discrete group of isometries of S^3, and this group may be used as the deck transformations of a manifold M covered by S^3. M is called a lens space and has a metric induced from S^3. Prove that M is not Riemannian symmetric, even though its covering is.

Remark. Notice that if, instead of connectedness, one assumes that the components of a Riemannian symmetric space M are isometric, then M is still a homogeneous Riemannian symmetric space, so the two notions are now equivalent. It is not hard to think of examples of Riemannian manifolds for which this does not hold but which still admit symmetries f_m, but they are rather artificial and will not be missed.

15. *Complex projective spaces.* Let S^{2d+1} be a Riemannian sphere of radius r. Then we have the principal fibre bundle (S^{2d+1}, S^1, CP^d) (problem 3.12), and S^1 acts as isometries, so the normal subspace to the vertical space at a point defines a connexion H in this principal bundle (problem 22). If we define $d\pi \mid_{H_s}$ to be an isometry between $CP^d_{\pi(s)}$ and H_s, this defines a Riemannian metric on CP^d. The bundle maps which are isometries on S^{2d+1} (this includes all of the unitary transformations of C^{d+1} restricted to S^{2d+1}) form a transitive group of isometries on CP^d, so CP^d is a Riemannian homogeneous space. A closer examination of the available bundle maps show that CP^d is also symmetric.

16. *Complex manifolds* [23, 30, 92]. Let M be a complex manifold of complex dimension d (see problems 1.7 and 3.10). M is called an *Hermitian manifold* if for each $m \in M$ there is given a positive definite Hermitian bilinear form $\langle \, , \, \rangle$ on $\mathscr{H}_m$, and this assignment is C^∞ in the sense that if $z_1, \ldots, z_d$ is a complex coordinate system at m, then the functions $g_{ij} = \langle \partial/\partial z_i, \partial/\partial z_j \rangle$ are complex-valued C^∞ functions, that is, $g_{ij} \in \mathscr{F}$. We cannot demand that $\langle \, , \, \rangle$ be holomorphic, since $g_{ij} = \bar{g}_{ji}$.

Problem 33. Since M is complex, its bundle of bases $B(M)$ is reduced to a bundle $CB(M)$ with group $GL(d, C)$, and $CB(M)$ may be viewed as the bundle of holomorphic bases of the holomorphic tangent spaces $\mathscr{H}_m$ to M. Show that the existence of a Hermitian structure on M is equivalent to the reduction of $CB(M)$ to a bundle $CF(M)$ with group $U(d) \subset GL(d, C)$. Hence or otherwise show that a Hermitian structure always exists. Show that $CF(M)$ may be taken to consist of holomorphic bases which are orthonormal with respect to the Hermitian forms.

Problem 34. Establish that $\langle \, , \, \rangle$ can be uniquely extended to a positive definite Hermitian form on $\mathscr{T}_m$ in such a way that

$$\langle s, t \rangle = \overline{\langle \bar{s}, \bar{t} \rangle} \qquad \text{if} \qquad s, t \in \mathscr{H}_m, \qquad \text{and} \qquad \langle \mathscr{H}_m, \overline{\mathscr{H}}_m \rangle = 0.$$

Problem 35. Show that if $s, t \in \mathscr{M}_m$, then

$$\langle s, t \rangle = 2R \langle Ps, Pt \rangle$$
$$= \tfrac{1}{2} R \langle \bar{s}, \bar{t} \rangle,$$

where $\bar{s} = s - iJs$, $\bar{t} = t - iJt$ and $R =$ "real part of".

This then defines a Riemannian structure on the underlying real manifold of M, and so $F(M)$ is defined and has a Riemannian connexion. Actually, we shall consider the $F(M)$ defined by the metric $\langle \, , \, \rangle' = 2\langle \, , \, \rangle$.

Show that J is an orthogonal transformation field with respect to this symmetric inner product.

Problem 36. Since $\mathscr{T}_m = \mathscr{M}_m + i\mathscr{M}_m$, any $t \in \mathscr{H}_m$ may be decomposed as $t = Rt + iIt$, with $Rt, It \in \mathscr{M}_m$. Show that $\langle Rt, It \rangle = 0$ and hence show that the map

$$i : CF(M) \to F(M)$$

given by $i(m, t_1, ..., t_d) = (m, Rt_1, ..., Rt_d, It_1, ..., It_d)$ is a well-defined imbedding of the real manifold $CF(M)$ and is consistent with the previously defined inclusion of $CB(M)$ in $B(M)$. We shall write $CF(M)$ for $i(CF(M))$. The following problems show that the Riemannian connexion on $F(M)$ does not in general reduce to a connexion on $CF(M)$.

Problem 37. Prove that the following assertions are equivalent:

(a) The Riemannian connexion on $F(M)$ reduces to a connexion on $CF(M)$, that is, if $b \in CF(M)$, then H_b is tangent to $CF(M)$.

(b) The parallel translate of a holomorphic tangent is holomorphic.

(c) Parallel translation commutes with the projection P. (See p. 51.)

(d) Parallel translation commutes with J.

(e) Covariant derivatives of J vanish.

(f) If X, Y are vector fields on M, then $\nabla_X(JY) = J\nabla_X(Y)$.

Problem 38. For $s, t \in \mathcal{M}_m$, define $\Omega(s, t) = -\langle Js, t\rangle$. Show that Ω is a differential 2-form on M and that

$$\Omega(s, t) = 2I\langle Ps, Pt\rangle$$
$$= \tfrac{1}{2}I\langle \tilde{s}, \tilde{t}\rangle.$$

M is said to be a *Kähler manifold* if $d\Omega = 0$. Prove that the following assertions are equivalent:

(a) $d\Omega = 0$.

(b) All covariant derivatives of Ω are zero.

(c) $\nabla_X(JY) = J\nabla_X(Y)$, X, Y any vector fields.

[*Hint*: prove the following implications:

$$\text{(a)} \Rightarrow \text{(c)} \Rightarrow \text{(b)} \Rightarrow \text{(a)}.$$

Recall problems 6.13, 6.14 and the vector field formula for the exterior derivative.]

Problem 39. A direct proof that M is Kähler if and only if the Riemannian connexion on $F(M)$ reduces to $CF(M)$ is outlined in this problem.

Let ϕ_{ij}, ω_k be the components of the connexion and solder forms on $F(M)$. Let $\tilde{\Omega} = \Omega \circ d\pi$. Define a linear transformation field $\tilde{J}$ on

$F(M)$ by: $\tilde{J}\,|_{V_b} = $ identity and $\tilde{J}\,|_{H_b} = (d\pi\,|_{H_b})^{-1} \circ J \circ d\pi$. Prove the following:

(a) $d\tilde{\Omega} = 0$ if and only if $d\Omega = 0$.

(b) The vertical tangent space to $CF(M)$ is spanned by the vector fields $F^{ij} + F^{i+d,j+d}$ and $F^{i,j+d} - F^{i+d,j}$.

On $CF(M)$ we have

(c) $\qquad \omega_i \circ \tilde{J} = -\omega_{i+d}\,,\ \omega_{i+d} \circ \tilde{J} = \omega_i \qquad (1 \leqslant i \leqslant d)$.

(d) $$\tilde{\Omega} = 2 \sum_{1 \leqslant i \leqslant d} \omega_i \omega_{i+d}.$$

(e) $\frac{1}{2} d\tilde{\Omega} = \displaystyle\sum_{i,j \leqslant d} (\omega_{ij} - \omega_{i+d,j+d})\,\omega_i \omega_{j+d}$

$$+ \sum_{i < j \leqslant d} (\omega_{i+d,j} + \omega_{i,j+d})(\omega_{i+d}\omega_{j+d} + \omega_i \omega_j).$$

(f) t tangent to $CF(M)$ implies Vt tangent to $CF(M) \leftrightarrow d\tilde{\Omega} = 0$.

Problem 40. Show that Ω is a form of type $(1, 1)$ and that $\Omega^d \neq 0$. If M is compact and Kähler, then show also that Ω^p is not exact for any $p \leqslant d$. (A form θ is *exact* if there is a form ψ such that $d\psi = \theta$.)

Problem 41. If $\langle\,,\,\rangle$ is a Riemannian form on a complex manifold with respect to which J is orthogonal, then construct a Hermitian form on M from which the Riemannian form is derived as in problem 35.

Problem 42. Show that a submanifold (complex) of a Kähler manifold is again Kähler.

Problem 48. *Complex projective space.* Let $t_0\,,\,...,\,t_d$ be homogeneous complex-valued coordinate functions on CP^d. Then a basis of coordinate systems may be described as follows. Let

$$U_i = \{p \in CP^d \mid t_i(p) \neq 0\}, \qquad i = 0, ..., d,$$

and define coordinate functions $\phi_i : U_i \to C^d$ by

$$z_j^i = z_j \circ \phi_i = t_j/t_i.$$

The ϕ_i then define a complex structure on CP^d. On each U_i consider the function $f_i = \Sigma_{j=0}^d z_j^i \bar{z}_j^i$. Notice that on $U_i \cap U_k$

$$f_k = f_i z_i^k \bar{z}_i^k.$$

Use this to show that there exists a real closed form Ω of type $(1, 1)$ on CP^d such that on U_j

$$\Omega = -i\,d'd''f_j. \quad \text{(Problem 4.28.)}$$

Show that the symmetric form associated with Ω is positive definite and admits J as an orthogonal transformation, and hence show that CP^d is a Kähler manifold with respect to the corresponding Hermitian structure.

A result of the preceding two problems is that every nonsingular projective variety admits a Kähler structure.

CHAPTER 8

Geodesics and Complete
Riemannian Manifolds

In the first section the local minimizing properties of geodesics are established with the help of the Gauss lemma, while in the second the Hopf-Rinow theorem on complete Riemannian manifolds is proved. In particular, it is shown that in the complete case geodesics realize global distances. The arc length of continuous curves is also discussed [*24, 33, 50, 83*].

8.1 Geodesics

In 6.3 we defined for any $m \in M$, $b = (m, e_1, ..., e_d) \in B(M)$, maps $\exp_m : M_m \to M$, $\overline{\exp}_b : M_m \to B(M)$, depending on a connexion on $B(M)$ and such that $\overline{\exp}_b$ was a rather natural lifting of $\exp_m$. We noticed that these maps in general were only defined on a neighborhood of the origin of M_m, they were C^∞ on this neighborhood, and $\exp_m$ was a diffeomorphism of a possibly smaller neighborhood of the origin. In order for them to be globally defined we must assume that geodesics from m are infinitely extendible.

We now assume M is a Riemannian manifold and we obtain $\overline{\exp}_b$ as follows:

Let $b = (m, f_1, ..., f_d) \in F(M)$. Let ϕ be the Riemannian connexion on $F(M)$, and let $p = \sum p_i f_i \in M_m$, and let $\bar{\sigma}$ be the unique integral curve of $\sum p_i E_i$ with $\bar{\sigma}(0) = b$. Then we define $\overline{\exp}_b(p) = \bar{\sigma}(1)$.

E_i is the restriction of the vector field E_i on $B(M)$ (6.1.2) to $F(M)$. $\overline{\exp}_b$ is C^∞ and is in fact the radially horizontal lifting of $\exp_m$ to $F(M)$. In particular, the image of M_m under $\overline{\exp}_b$ is in $F(M)$.

If Φ is the curvature form of the Riemannian connexion, ω the solder form on $F(M)$ as defined in 7.3, then we define forms θ, Θ, ψ on M_m as follows:

$$\theta = \overline{\exp_b}{}^* \phi$$

$$\Theta = \overline{\exp_b}{}^* \Phi$$

$$\psi = \overline{\exp_b}{}^* \omega,$$

and the real-valued forms θ_{ij}, Θ_{ij}, ψ_i are defined accordingly. The structural equations become

$$d\psi = -\theta\psi$$

$$d\theta = -\tfrac{1}{2}[\theta, \theta] + \Theta.$$

(In terms of components, the structural equations become

$$d\psi_i = \sum_{k=1}^{d} \theta_{ki}\psi_k$$

$$d\theta_{ij} = -\sum_{k=1}^{d} \theta_{ik}\theta_{kj} + \Theta_{ij}.)$$

If $p \in M_m$, $\rho = $ ray from m to p, $\sigma = \exp_m \circ \rho$, $\overline{\exp_b}p = (m, f_1', ..., f_d')$ (so f_i' is the parallel translate of f_i along σ), then if s, $t \in (M_m)_p$,

(a) $d \exp_m s = \sum \psi_i(s) f_i'$

(b) $\begin{cases} \langle d \exp_m s, d \exp_m t \rangle = \langle \psi(s), \psi(t) \rangle = \sum \psi_i(s)\, \psi_i(t), \\ \| d \exp_m s \|^2 = \| \psi(s) \|^2 = \sum \psi_i(s)^2. \end{cases}$

[(b) is an immediate consequence of (a), which in turn follows easily from the definition of ψ.]

Thus, if $x_1, ..., x_d$ is the dual base to the f_i, $s = \sum s_i D_{x_i}(p)$, then $\psi_i(s) - s_i$ is a measure of the difference between parallel translation in M and M_m.

If $p = \sum p_i f_i \in M_m$, $s = c \sum p_i D_{x_i}(p) \in (M_m)_p$, that is, s is tangent to the ray ρ above, then

(c) $\psi_i(s) = cp_i$

(d) $\theta(s) = 0.$

This says that the lengths of radial vectors are preserved under $d \exp_m$ and that tangents to horizontal curves in $F(M)$ are horizontal tangents.

A C^∞ *rectangle* in M is a map Q of a rectangle $[a, b] \times [c, d]$ in R^2 into M which can be extended to a C^∞ map of a neighborhood of the rectangle. $Q(a, c)$ is the *initial corner* of Q, while the *base* of Q is the

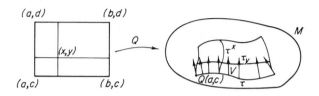

FIG. 28.

curve τ defined by $Q \circ j_c$, where $j_c(t) = (t, c)$. More generally, the curves $\tau_y = Q \circ j_y$ are called *longitudinal*, while the curves $\tau^x = Q \circ {}'j_x \, ({}'j_x(t) = (x, t))$ are *transversal*. The "vector field" V defined along the base τ by $V(s) = \tau^s{}_*(c)$ is the *associated vector field of Q* (it is actually a curve in the tangent bundle of M), and Q is an *associated rectangle* of V.

Canonical lifting. If Q is a C^∞ rectangle in a Riemannian manifold M, and if $F(M)$ has the Riemannian connexion ϕ, then for every $f \in \pi^{-1}(Q(a, c))$ there exists a unique C^∞ rectangle $\tilde{Q}$ in $F(M)$ with initial corner f such that

(a) $Q = \pi \circ \tilde{Q}$, that is, $\tilde{Q}$ is a lifting of Q,

(b) $\phi(\tilde{Q} \circ j_{y*}) = 0$, that is, the longitudinal curves of $\tilde{Q}$ are horizontal, and

(c) $\phi(\tilde{Q} \circ {}'j_{a*}) = 0$, that is, the initial transversal curve is horizontal.

Problem 1. Prove the existence, uniqueness, and differentiability of the canonical lifting of a C^∞ rectangle.

Theorem 1 (Gauss' lemma). If Q is a C^∞ rectangle in M, $Q :$ $[a, b] \times [c, d] \to M$, whose longitudinal curves are geodesics and the tangents to these geodesics are all of the same length, then if V is the associated vector field, the function $\langle \tau_{c*}, V \rangle$ is a constant. In particular, if $\tau_{c*}(a) \perp V(a)$, then $\tau_{c*}(t) \perp V(t)$ for all $t \in [a, b]$.

Proof.   Let $\underset{\sim}{Q}$ be the canonical lifting to $f \in F(M)$ of Q. Also, write $\phi^Q = \underset{\sim}{Q}{}^*\phi$, $\omega^Q = \underset{\sim}{Q}{}^*\omega$. The first structural equation becomes $d\omega^Q = -\phi^Q \omega^Q$, which when applied to (D_1, D_2) gives

$$D_1\omega^Q(D_2) - D_2\omega^Q(D_1) - \omega^Q([D_1, D_2]) = -\phi^Q(D_1)\,\omega^Q(D_2) + \phi^Q(D_2)\omega^Q(D_1).$$

Now $[D_1, D_2] = 0$, and $\phi^Q(D_1) = 0$ since the longitudinal curves of $\underset{\sim}{Q}$ are horizontal; so we have, taking inner products with $\omega^Q(D_1)$,

$$\langle D_1\omega^Q(D_2),\, \omega^Q(D_1)\rangle - \langle D_2\omega^Q(D_1),\, \omega^Q(D_1)\rangle = \langle \phi^Q(D_2)\,\omega^Q(D_1),\, \omega^Q(D_1)\rangle$$
$$= 0$$

since $\phi^Q(D_2)$ is skew-symmetric.

Now $\omega^Q(D_1) \circ j_y$ is constant since the curve τ_y is a geodesic, so

$$D_1 \langle \omega^Q(D_2),\, \omega^Q(D_1)\rangle = \langle D_1\omega^Q(D_2),\, \omega^Q(D_1)\rangle;$$

also,

$$D_2 \langle \omega^Q(D_1),\, \omega^Q(D_1)\rangle = 2\, \langle D_2\omega^Q(D_1),\, \omega^Q(D_1)\rangle.$$

But $\langle \omega^Q(D_1),\, \omega^Q(D_1)\rangle$ is constant, since we assumed the tangents to the longitudinal curves all had the same length, so

$$D_2 \langle \omega^Q(D_1),\, \omega^Q(D_1)\rangle = 0.$$

Hence, the above equation becomes $D_1 \langle \omega^Q(D_2),\, \omega^Q(D_1)\rangle = 0$. But clearly, along the base curve $\langle \omega^Q(D_2),\, \omega^Q(D_1)\rangle = \langle V, \tau_{c*}\rangle$. QED

Corollary.   Let $p \in M_m$, ρ = ray from 0 to p, $\sigma = \exp_m \circ \rho$, and $s \in (M_m)_p$. Then $s \perp \rho$ (in the Euclidean inner product) implies that $d \exp_m s \perp \sigma$.

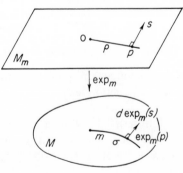

Fig. 29.

This follows by applying Gauss' lemma to a rectangle Q whose initial transversal is the degenerate curve $m \in M$ and the longitudinal curves are the images under $\exp_m$ of the rays from 0 in M_m. In particular, the base is σ. The exact construction is left to the reader. (This rectangle may be described briefly as "a piece of pie.")

The following result expresses the fact that locally geodesics minimize arc length among broken C^∞ curves.

Theorem 2.   Let $B \subset M_m$ be a ball about 0 on which $\exp_m$ is a diffeomorphism, let $p \in B$, $\rho =$ ray from 0 to p, $\sigma = \exp_m \circ \rho$, and let τ be any broken C^∞ curve from m to $\exp_m p$ in M. Then $|\tau| \geqslant |\sigma|$, and equality holds only if τ is a broken C^∞ reparametrization of σ.

Proof.   Let $x_1, ..., x_d$ be dual to $f_1, ..., f_d$ on M_m and define the following objects on either B or $B' = \exp_m B$:

$$r = \left(\sum x_i^2 \right)^{1/2}, \qquad \bar{r} = r \circ \exp_m^{-1}$$

$$T = \sum \frac{x_i D_{x_i}}{r}, \text{ the radial unit vector field defined on } B - \{0\}.$$

$$\bar{T} = d \exp_m T \circ \exp_m^{-1}, \text{ defined on } B' - \{m\}.$$

If $q \in M_m$, $s \in (M_m)_q$, then we write $s = s_T + s_N$ where s_T is a multiple of $T(q)$ and $s_N \perp T(q)$. For $t \in M_b$, $b \in B'$, we write similarly $t = t_T + t_N$, where t_T is a multiple of $\bar{T}(b)$ and $t_N \perp \bar{T}(b)$. We have $d \exp_m s = d \exp_m s_T + d \exp_m s_N$, and from the corollary above, $d \exp_m s_N \perp d \exp_m s_T$. Hence,

(i) $d \exp_m s_T = (d \exp_m s)_T$. Also, from (b), p. 146,

(ii) $\| d \exp_m s_T \| = \| s_T \|$.

Let $[a, b]$ be the interval on which τ is defined, and let $c \in [a, b]$ be the smallest number such that $\bar{r}(\tau(c)) = r(p) = |\sigma|$. Now define a curve η on $[a, c]$ by

$$\eta = \sigma \circ \frac{\bar{r}}{\| p \|} \circ \tau.$$

Noting that $\| \sigma_* \| = \| p \|$, we have

$$\eta_* = d\eta \left(\frac{d}{du}\right) = d\sigma \circ \frac{dr}{\| p \|} \circ d\exp_m^{-1} \circ d\tau \left(\frac{d}{du}\right)$$

$$= \frac{1}{\| p \|} \, d\sigma \circ dr \circ d\exp_m^{-1}(\tau_*)$$

$$= \frac{1}{\| p \|} \, d\sigma \left(\| (d\exp_m^{-1}\tau_*)_T \| \frac{d}{du}\right)$$

$$= \frac{1}{\| p \|} \, \| (d\exp_m^{-1}\tau_*)_T \| \, \sigma_* \, ;$$

hence,

$$\| \eta_* \| = \| (d\exp_m^{-1}\tau_*)_T \| = \| d\exp_m(d\exp_m^{-1}\tau_*)_T \| \qquad \text{by (ii)}$$

$$= \| \tau_{*T} \| \qquad \text{by (i);}$$

this is true on $[a, c]$. Therefore, if $\tau' = \tau \, |_{[a,c]}$, we have

$$| \tau | \geqslant | \tau' | \geqslant | \eta | \geqslant | \sigma |,$$

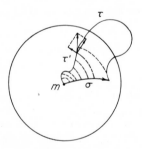

Fig. 30.

as asserted. If $| \tau | = | \sigma |$, then we clearly must have $\tau_*(u) = 0$ for $u > c$ and $\tau_{*N} = 0$, from which it follows that τ and σ have the same image. QED

Corollary 1. The square of the distance to m, $\rho(m, I)^2$, is a C^∞ function on B'.

Proof. It equals $\Sigma (x_i \circ \exp_m^{-1})^2$. ($I$ is the identity map on M.)

Corollary 2. If $B(m, c)$ denotes the ball of radius c about m, then for $c <$ radius of B, $\exp_m(B(0, c)) = B(m, c)$.

Corollary 3. If τ is a broken C^∞ curve from m to n such that $|\tau| = \rho(m, n)$, then τ is a broken C^∞ reparametrization of a geodesic.

Proof. τ minimizes arc length from m to n, so τ locally minimizes arc length and, by the theorem, is locally a broken C^∞ reparametrization of a geodesic. This is enough. QED

Lemma 1. Let $m \in M$, O a ball $\subset M_m$, such that $\exp_m$ is a diffeomorphism : $O \to U$, $\gamma : (a, b) \to U$ a C^∞ curve in U. Suppose that $r = \rho(m, \gamma)$ has an argument t such that $r'(t) = 0$. Then the geodesic in U from m to $\gamma(t)$ is perpendicular to $\gamma_*(t)$ at $\gamma(t)$.

Proof. This is an immediate consequence of Gauss' lemma applied to the lifting of the curves to M_m and the comparable fact for Euclidean space.

Theorem 3. Let N, P be submanifolds of M, σ a geodesic from $n \in N$ to $p \in P$ such that $|\sigma| = \rho(P, N)$. Then σ is perpendicular to both N and P.

Proof. It is obvious that a piece of σ minimizes arc length from N or P to any point on σ. For such points which are sufficiently close to N or P lemma 1 then shows that σ is perpendicular to curves in N or P which pass through n or p, respectively, and hence σ is perpendicular to N_n and P_p . QED

If N is a submanifold of a Riemannian manifold M, let $\perp(N)$, the *normal bundle* to N, be defined by

$$\perp(N) = \{(n, t) \in T(M) \mid t \in M_n \text{ for some } n \in N \text{ and } t \perp N_n\}$$

[cf. 3.3(4)].

Problem 2. Show that $\perp(N)$ is a submanifold of $T(M)$ and that $\text{Exp}\mid_{\perp(N)}$ is nonsingular on the trivial cross section of $\perp(N)$.

The *tubular neighborhood*, $\perp_r(N)$, of N with radius r in $\perp(N)$ is the open set of $\perp(N)$ which intersects each fibre in the open ball of radius r about the origin of the fibre.

Theorem 4. If N is a compact submanifold of the Riemannian manifold M, then there exists an $r > 0$ such that $\exp$ maps $\perp_r(N)$

diffeomorphically. The image of $\perp_r(N)$, called a *tubular neighborhood* of N in M, has the property that all its points are joined to N by unique geodesics which minimize arc length to N.

The proof is left as an exercise.

A local version of this gives us the existence of normal coordinates for N in the following sense.

Let $n \in N$. A coordinate system at n in M is *normal for N* if the points of N correspond to part of a linear subspace of dimension equal to that of N and the straight lines perpendicular to this subspace correspond to geodesics perpendicular to N.

Problem 3. If $i : N \to M$ is an immersion (di is an isomorphism into for every point), define a normal bundle of the immersion and tubular neighborhoods in it. When N is compact show that Exp $: T(M) \to M$ gives rise to an immersion of $\perp_r(N)$ into M, some r. Give an example to show that the property of having unique minimizing geodesics to $i(N)$ need not hold for points of the immersion of $\perp_r(N)$.

Problem 4. Consider the flat 2-dimensional torus obtained by identifying opposite sides of the unit square in R^2 having opposite corners $(0, 0)$ and $(1, 1)$. Sketch the locus of points at distance $2/3$ from the corner point.

If we obtain a noncomplete (see below) manifold from this torus by removing the closed line segment from $(1/4, 0)$ to $(1/4, 1/2)$, how does this change the locus?

Problem 5. Show that a Riemannian covering map is distance-decreasing.

8.2 Complete Riemannian Manifolds

Assume M is a Riemannian manifold. We have a map $E : T(M) \to M \times M$, given by: $E(m, t) = (m, \exp_m t)$. For E to be defined on all of $T(M)$ we must assume that all geodesics are infinitely extendible. But in any case, E is defined on a neighborhood of the zero cross section of $T(M)$ and is in fact C^∞ there. We also have

Lemma 2. For each $m \in M$, dE is an isomorphism on $T(M)_{(m,0)}$, so by the inverse function theorem, E is a diffeomorphism of a neighborhood of $(m, 0)$ onto a neighborhood of (m, m).

Proof. It is sufficient to prove that dE maps $T(M)_{(m,0)}$ onto $M \times M_{(m,m)}$, since the dimensions are the same. Let $\pi_i : M \times M \to M$ be the projection onto the ith factor $i = 1$, 2. Now we know that $E \mid_{\pi^{-1}(m)}$, which is given by $E \mid_{\pi^{-1}(m)} (m, t) = (m, \exp_m t)$, maps $(M_m)_0$ onto the tangent space to $\pi_1^{-1}(m)$ at (m, m). We conclude by showing that dE maps the tangents to the zero cross section of $T(M)$ onto the tangent space to the diagonal of $M \times M$, which suffices since $M \times M_{(m,m)}$ is clearly spanned by tangents to $\pi_1^{-1}(m)$ and tangents to the diagonal.

Let $D : M \to M \times M$ be the diagonal map, $D(m) = (m, m)$. Then we have $E \mid_{\text{(zero cross section)}} = D \circ \pi \mid_{\text{(zero cross section)}}$. $d\pi$ is onto, and dD is onto the tangent space to the diagonal. QED

Lemma 3. Let C be any compact subset of M. Then there exists a $c > 0$ such that, for each $m \in C$, $\exp_m$ is defined on $B(O(m), c)$ and maps it diffeomorphically onto $B(m, c)$, where $O(m)$ is the origin in M_m.

Proof. We first note that if all geodesics are not infinitely extendible, $\exp_m$ may only be defined in a neighborhood of $O(m)$. However, by the theory of differential equations, it is clear that for each $m \in M$ $\exp_m$ is defined on a ball whose radius is a continuous function of m, and hence we may take a c_1 such that $\exp_m$ is defined on $B(O(m), c_1)$ for every $m \in C$. By corollary 2 to theorem 2, we need only show that there exists $c > 0$ such that, for each $m \in C$, $\exp_m$ is a diffeomorphism on $B(O(m), c)$. Lemma 2 essentially says this locally, and hence by compactness the result follows.

However, we must first translate lemma 2, which says precisely that for every $m \in C$, there exists a neighborhood P_m of $(m, 0)$ which is mapped diffeomorphically onto a neighborhood of (m, m). Therefore, for every $n \in \pi (P_m)$, $d \exp_n$ is regular (that is, $1 - 1$ onto) and hence $\exp_n$ is a diffeomorphism on the set $\{t \in M_n \mid (n, t) \in P_m\} = P_{m,n}$; so we wish to show there is a $c_m > 0$ such that the ball, $B(O(n), c_m)$, in the Riemannian metric is contained in $P_{m,n}$, for all $n \in \pi(P_m)$.

Using the facts that $T(M)$ has a local product structure with the topology of the fibre being given by the Euclidean metric, that the Riemannian and Euclidean metrics are equivalent on each tangent space (see lemma 7.1), and that the Riemannian metric is continuous, one finds that P_m contains a neighborhood of the form

$$P'_m = \{(n, t) \mid n \in U_m, \| t \| < c_m\},$$

where U_m is a neighborhood of m and $c_m > 0$. Hence for $n \in U_m$, $\exp_n$ maps $B(O(n),\ c_m)$ diffeomorphically onto $B(n, c_m)$. By compactness, C is covered by a finite number of U_m, to each of which corresponds a c_m. Letting $c = \min c_m$ gives the desired result. QED

It now follows that the square of the distance function, ρ^2, is C^∞ on a neighborhood of (m, m) in $M \times M$.

Theorem 5. (Hopf-Rinow [40, 77, 93]). Consider the following conditions on a connected Riemannian manifold M:

(a) M is complete.

(b) All bounded closed subsets of M are compact.

(c) For some point m of M, all geodesics from m are infinitely extendible.

(d) All geodesics are infinitely extendible.

(e) Any m, $n \in M$ can be joined by a geodesic whose arc length equals $\rho(m, n)$.

The conditions (a)-(d) are equivalent, and they imply (e).

Problem 6. Find an example to show that (e) does not imply (a). Also find an example to show that a minimizing geodesic between two points need not be unique; in fact, there may be infinitely many.

Notice that (c) can be stated "$\exp_m$ is defined on all of M_m", and (d) can be stated "The Riemannian connexion is complete" (6.3).

Proof. In any metric space it is true that (b) implies (a). That (d) implies (c) is trivial, and that (a) implies (d) follows from extendibility theorems in differential equations (see appendix).

We therefore have only to prove the implications from (c) to (b) and from (b) to (e).

We now fix $m_0 \in M$, and define, for any real $r > 0$,

$$\bar{B}_r = \{m \in M \mid \rho(m_0, m) \leqslant r\} = \overline{B(m_0, r)};$$

$$E_r = \{m \in \bar{B}_r \mid m \text{ can be joined to } m_0 \text{ by a geodesic of length } \rho(m, m_0)\}.$$

It suffices, assuming (c) for m_0, to prove (1) E_r is compact and (2) $E_r = \bar{B}_r$; for any bounded subset is in a $\bar{B}_r$, so by (1) and (2) a closed bounded subset is compact, which is (b); then since (b) implies (d), we can use any m for m_0, so that (2) implies (e).

Proof of (1). Let $\bar{E}_r = \{p \in M_{m_0} \mid \rho(m_0, \exp_{m_0}p) = \| p \| \leqslant r\}$, and define $f(p) = \rho(m_0, \exp_{m_0}p) - \| p \|$. Then we have

$$\bar{E}_r = f^{-1}(0) \cap \overline{B(O(m_0), r)},$$

which is clearly compact, since it is closed and bounded. Compactness of E_r then follows from the fact that $E_r = \exp_{m_0}\bar{E}_r$ and the continuity of $\exp_{m_0}$. QED

For the proof of (2) we need the following:

Lemma 4. If $E_r = \bar{B}_r$ for some r, and if $\rho(m_0, n) > r$, then there exists m such that $\rho(m_0, m) = r$ and $\rho(m_0, n) = r + \rho(m, n)$.

Proof. For $k = 1, 2, ...$, choose σ_k, a broken C^∞ curve from m_0 to n with $| \sigma_k | < \rho(m_0, n) + 1/k$ (this is possible by the definition of ρ). Let m_k be the last point on σ_k in $\bar{B}_r$, so $\rho(m_0, m_k) = r$. By com-

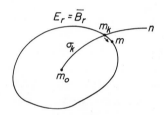

FIG. 31.

pactness of $E_r = \bar{B}_r$, the m_k have a limit point m, and by passing to a subsequence we may assume $\{m_k\}$ converges to m. Now

$$\rho(m_0, n) \leqslant \rho(m_0, m) + \rho(m, n) = r + \rho(m, n).$$

On the other hand, $\rho(m_0, n) > | \sigma_k | - 1/k = |$ part of σ_k from m_0 to $m_k | + |$ part of σ_k from m_k to $n | -1/k \geqslant r + \rho(m_k, n) - 1/k$, which has limit $r + \rho(m, n)$, so $\rho(m_0, n) \geqslant r + \rho(m, n)$, so equality holds. This proves lemma 4.

We now prove (2) using the connectedness of the nonnegative real numbers; that is, we show:

(i) $E_0 = \bar{B}_0$.
(ii) $E_r = \bar{B}_r$, $r' < r$, implies $E_{r'} = \bar{B}_{r'}$.
(iii) $E_{r'} = \bar{B}_{r'}$ for all $r' < r$ implies $E_r = \bar{B}_r$.

(iv) $E_r = \bar{B}_r$ implies there exists $c > 0$ such that
$$E_{r+c} = \bar{B}_{r+c}.$$

(i) and (ii) are trivially true.

Proof of (iii).   Let $m \in \bar{B}_r$. If $m \in \bar{B}_{r'}$ for $r' < r$, then by hypothesis $m \in E_{r'} \subset E_r$. Hence, we assume $\rho(m_0, m) = r$. By lemma 4 we may choose a sequence $\{m_k\}$ which has limit m, where $m_k \in \bar{B}_{r_k}$, $r_k < r$. Then by hypothesis, $m_k \in E_{r_k} \subset E_r$, so $m \in E_r$ by compactness of E_r, which proves (iii).

Proof of (iv).   By lemma 3, there exists $c > 0$ such that for each $m \in \bar{B}_r$, $\exp_m$ maps $B(O(m), 2c)$ diffeomorphically onto $B(m, 2c)$, since E_r is compact. Let $n \in \bar{B}_{r+c}$. We show $n \in E_{r+c}$. By lemma 4 there exists $m \in \bar{B}_r$ such that $\rho(m_0, n) = r + \rho(m, n)$, where $\rho(m_0, m) = r$. Therefore, $\rho(m, n) \leqslant c$, so there exists a geodesic γ from m to n with $|\gamma| = \rho(m, n)$. Let σ be a geodesic from m_0 to m with $|\sigma| = \rho(m_0, m) = r$. Then $\sigma + \gamma$ is a broken C^∞ curve from m_0 to n with $|\sigma + \gamma| = |\sigma| + |\gamma| = r + \rho(m, n) = \rho(m_0, n)$, so by corollary 3 to theorem 2, $\sigma + \gamma$ can be reparametrized as a geodesic. Hence, $n \in E_{r+c}$.

This completes the proof of the theorem.

It has been shown by K. Nomizu and H. Ozeki [67] that every connected paracompact manifold admits a complete Riemannian metric; furthermore, they show that if every Riemannian metric is complete then the manifold is compact. The converse, that a compact metric space is complete, is well known.

Problem 7.   Construct a noncomplete connected Riemannian manifold of infinite diameter such that no two points at distance greater than 1 from each other may be connected by a curve which minimizes arc length.

Problem 8.   Let $i : N \to M$ be an isometric imbedding. Let ρ be the Riemannian distance on M, $\rho' = \rho \circ i$, ρ'' the Riemannian distance on N. Give examples to show that all eight possibilities for these metrics to be complete or not may occur.

Problem 9.   Show that for a connected Lie group which admits a left and a right invariant metric the exponential map is onto.

Problem 10.   Let M and N be complete Riemannian manifolds. Show that the Riemannian product $M \times N$ is complete. Completeness of affine connections is also a property preserved under products.

Problem 11.   Let M be a complete Riemannian manifold.

(a) Then any Riemannian covering of M is complete.

(b) For every $m \in M$ there is a geodesic segment starting and ending at m in every homotopy class of loops at m.

Problem 12.   Give examples to show that the result in problem 11(b) may or may not hold if M is not complete.

Problem 13.   Let M be a complete Riemannian manifold which is not simply connected. Show that $\rho(m, I)^2$ cannot be a C^∞ function on all of M.

Example.   Consider a mechanical system with a finite number of degrees of freedom and no elements which dissipate energy. Then the *configuration space* is a manifold M which is a mathematical model of the totality of all positions of the system. A curve in M is generally thought of as occurring in time, so that a tangent vector is an assignment of velocities to the elements of the system consistent with the constraints of the system. The *phase space* of the system is $T(M)$. The kinetic energy given by an assignment of velocities is a positive definite quadratic form on each M_m, and thus gives a Riemannian metric on M.

A *force field* on M is a 1-form θ; the integral of this 1-form on a curve is the amount of work done in traversing the curve. If it is a *conservative* force field then $\theta = -dV$, where V is the potential energy.

If D is the covariant derivative symbol of the metric and X is the vector field defined by $2\langle X, Y \rangle = \theta(Y)$ for every Y, then Newton's laws of motion for the system become

$$D_{\gamma_*}\gamma_* = X,$$

where γ is a curve parametrized by time.

In particular, when the motion is free the path of the motion is a geodesic and the time is proportional to the length of the geodesic. The Riemannian manifold is thus complete if the system will coast indefinitely when given an arbitrary push. In this case the system may be started in any one configuration with kinetic energy 1 in such a way that it attains another given configuration by free motion with an elapse of time equal to the distance between the two points on M.

More specifically, the configuration space of a rigid body constrained to rotate about its center of gravity is homeomorphic to $P^3 = SO(3)$. The kinetic energy metric is invariant if and only if the ellipsoid of inertia is spherical.

8.3 Continuous Curves

Let γ be a continuous curve, $\gamma : [a, b] \to M$, and let

$$a = t_0 < t_1 < t_2 < \cdots < t_p < b = t_{p+1}$$

be a sequence of numbers between a and b. Then we define the *arc length* of γ by $|\gamma| = \sup \{\Sigma_{i=0}^{p}\, \rho(\gamma(t_i), \gamma(t_{i+1})) \mid$ all such sequences $t_i\}$.

Proposition 1. If $\gamma : [a, b] \to M$ is a continuous curve from m to n with $|\gamma| = \rho(m, n)$, then γ is a continuous reparametrization of a geodesic. This implies that geodesics locally minimize arc length among continuous curves.

Proof. Pick a $c > 0$ by lemma 3 with respect to the image of γ. Take $a < t_1 < t_2 < b$ such that $\rho(\gamma(t_1), \gamma(t_2)) < c/2$, so there exists a geodesic σ from $\gamma(t_1)$ to $\gamma(t_2)$ with $|\sigma| = \rho(\gamma(t_1), \gamma(t_2))$. We claim that for every $t \in (t_1, t_2)$, $\gamma(t)$ lies on σ. For choose geodesics σ_1, σ_2 from $\gamma(t_1)$ to $\gamma(t)$, $\gamma(t)$ to $\gamma(t_2)$, respectively, with $|\sigma_1| = \rho(\gamma(t_1), \gamma(t))$ and $|\sigma_2| = \rho(\gamma(t), \gamma(t_2))$. Then since γ clearly locally minimizes arc length, we have $|\sigma_1 + \sigma_2| = \rho(\gamma(t_1), \gamma(t)) + \rho(\gamma(t), \gamma(t_2)) = |\gamma$ from $\gamma(t_1)$ to $\gamma(t_2)| + |\gamma$ from $\gamma(t)$ to $\gamma(t_2)| = \rho(\gamma(t_1), \gamma(t_2))$, so by the theorem 2, $|\sigma_1 + \sigma_2|$ is a broken C^∞ reparametrization of σ, which proves $\gamma(t)$ lies on the image of σ. Hence, γ is locally a continuous reparametrization of a geodesic; so in particular it is a reparametrization of a broken C^∞ curve. The result then follows from corollary 3 to theorem 2.

Proposition 2. If γ is a broken C^∞ curve, then the two definitions of arc length agree.

Proof. Let $\{\gamma\}$ be the length of γ as a continuous curve, retaining $|\gamma|$ as the notation for the integral of tangent lengths. Then it follows trivially from the definition of distance that $\{\gamma\} \leqslant |\gamma|$. We also have, easily, that both definitions are additive:

$$\{\gamma + \sigma\} = \{\gamma\} + \{\sigma\}, \qquad |\gamma + \sigma| = |\gamma| + |\sigma|.$$

Now suppose γ is a curve for which $\{\gamma\} + k = |\gamma|$, $k > 0$. Then by splitting γ in half, we must have that the discrepancy on one of the halves is at least $k/2$. By repeatedly halving we get a nested sequence of parameter values $s_n \leqslant u_n$ such that the discrepancy of γ restricted to $[s_n, u_n]$ is at least $k/2^n$, and $u_n - s_n = c/2^n$, where $c = b - a$, $[a, b]$ is the interval of definition of γ. Let t be the common limit of s_n and u_n, and let t_n be a number in $[s_n, u_n]$ such that

$$\| \gamma_*(t_n) \| \, c/2^n = \int_{s_n}^{u_n} \| \gamma_* \|,$$

which exists by the mean value theorem. Then

$$\| \gamma_*(t_n) \| \, c/2^n \geqslant k/2^n + \{\gamma \mid_{[s_n, u_n]}\} \geqslant k/2^n + \rho(\gamma(s_n), \gamma(t)) + \rho(\gamma(t), \gamma(u_n)).$$

Now reparametrize γ so that $t = 0$, multiply this inequality by $2^n/c$, and take limits, getting

$$\| \gamma_*(0) \| \geqslant k/c + \overline{\lim_{n \to \infty}} \frac{\rho(\gamma(s_n), \gamma(0)) + \rho(\gamma(0), \gamma(u_n))}{u_n - s_n}.$$

Let x_i be normal coordinates at $\gamma(0)$, and $f_i = x_i \circ \gamma$. Then we have $\rho(\gamma(s), \gamma(0)) = (\Sigma f_i(s)^2)^{1/2} = (\Sigma f_i'(\theta_i s)^2)^{1/2} |s|$, where $0 < \theta_i < 1$, by the mean value theorem. Thus if we write

$$g(s) = \rho(\gamma(s), \gamma(0))/|s|$$

and

$$g(u_n) \, u_n - g(s_n) \, s_n = g(u_n)(u_n - s_n) + (g(u_n) - g(s_n)) \, s_n,$$

we get

$$\frac{\rho(\gamma(s_n), \gamma(0)) + \rho(\gamma(0), \gamma(u_n))}{u_n - s_n} \leqslant g(u_n) + |g(u_n) - g(s_n)|.$$

Since $\lim_{n \to \infty} g(u_n) = \| \gamma_*(0) \|$, this shows

$$\| \gamma_*(0) \| \geqslant k/c + \| \gamma_*(0) \|,$$

which contradicts $k > 0$. QED

Problem 14. Let $\gamma : [a, b] \to M$ be a continuous curve with finite length in a Riemannian manifold. Show that γ can be uniformly approximated by broken geodesics.

Problem 15. Let $\phi : M \to N$ be a map between Riemannian manifolds which is onto and preserves distances. Prove that ϕ is an isometry, (See [50], p. 169.) This says that the topological metric of a Riemannian manifold determines both the Riemannian structure and the differential structure.

Problem 16. Let $\phi : M \to N$ be a C^∞ map of complete Riemannian manifold M onto Riemannian manifold N such that for every $m \in M$, M_m decomposes orthogonally into subspaces V_m and H_m, V and H C^∞ distributions, where $V_m = \ker(d\phi_m)$ and $d\phi_m$ restricts to an isometry from H_m onto $N_{\phi(m)}$. Show that:

(a) If γ is a C^∞ curve in N and $\phi(m) = \gamma(0)$, then there is a unique lift $\bar{\gamma}$ of γ to M such that $\bar{\gamma}(0) = m$, $\gamma = \phi \circ \bar{\gamma}$, and $|\gamma| = |\bar{\gamma}|$.

(b) The lift, as in (a), of a geodesic is a geodesic.

CHAPTER 9

Riemannian Curvature

The main properties of the Riemannian curvature are established, including a direct, but for the most part impractical, method of computing curvature. Following a selection of examples, the Jacobi equation is established for vector fields associated with rectangles with geodesic longitudinals and a number of local and global consequences are derived. In particular, it is shown that for a complete Riemannian manifold with nonpositive curvature, the exponential is a covering map [*24, 33, 50, 83*].

9.1 Riemannian Curvature

Let M be a d-dimensional Riemannian manifold with metric $\langle \, , \, \rangle$ and curvature transformation R_{st}, s, t tangents to M. A *plane section P at $m \in M$* is a 2-dimensional subspace of M_m.

Let P be a plane section at m, and let s, $t \in M_m$ be two vectors spanning P.

The *Riemannian* (or *sectional*) *curvature of P, K(P)*, is defined by

$$K(P) = \frac{\langle R_{st}s, t \rangle}{A(s, t)^2}$$

where $A(s, t) = (\| s \|^2 \| t \|^2 - \langle s, t \rangle^2)^{1/2}$ is the area of the parallelogram spanned by s and t.

The first aim is to prove that $K(P)$ depends on P alone and not on the particular choice of s and t spanning P. Simultaneously, it will be proved that $K(P)$ determines R_{st}, and so nothing is lost by considering the Riemannian curvature instead of the curvature form on $F(M)$.

161

Note that, if dim $M = 2$, there is only one plane section at each $m \in M$, and so K is a real-valued function on M, called the *Gaussian curvature*.

Problem 1. Let $f : M \to N$ be a local isometry. Show that f preserves curvature. Prove that the d-dimensional Riemannian sphere has constant curvature, and hence also the d-dimensional real projective space.

Lemma 1. For $x, y, z, w \in M_m$, the curvature tensor R satisfies the following properties:

(a) $R_{xy} = -R_{yx}$,   (b) $\langle R_{xy}z, w \rangle = -\langle z, R_{xy}w \rangle$,

(c) $R_{xy}z + R_{zx}y + R_{yz}x = 0$,   (d) $\langle R_{xy}z, w \rangle = \langle R_{zw}x, y \rangle$.

One way of interpreting these properties is to view R as a linear transformation of the second Grassmann space $G^2{}_m$, that is, the space of bivectors. To do this we define R_{xy} to be the bivector which satisfies $\langle R_{xy}, zw \rangle = \langle R_{xy}z, w \rangle$ for all decomposible bivectors zw. Then (b) says that R_{xy} is a well-defined bivector, (a) says that it depends only on the bivector xy and not on x and y individually, so that $xy \to R_{xy}$ can be extended linearly to an endomorphism of all bivectors; (d) says that R is a symmetric transformation of bivectors, so the corresponding quadratic form determines the transformation. Furthermore, (c) says that the quadratic form is determined by its values on the decomposible elements alone. (See corollary 2 below.) If $x_1, \ldots, x_d$ is a coordinate system at m, then the classical object R_{ijkl} is given by

$$R_{ijkl} = \langle R_{X_i X_j} X_k, X_l \rangle$$

where $X_i = D_{x_i}$. The above formulas then correspond to the classical ones, namely:

(a') $R_{ijkl} = -R_{jikl}$,   (b') $R_{ijkl} = -R_{ijlk}$,

(c') $R_{ijkl} + R_{kijl} + R_{jkil} = 0$,   (d') $R_{ijkl} = R_{klij}$.

Proof of lemma 1. Let $x, y \in M_m$, $b \in F(M)$, $\bar{x}, \bar{y} \in F(M)_b$ such that $d\pi \bar{x} = x$, $d\pi \bar{y} = y$. Then by 6.1.5, if $z \in M_m$,

$$R_{xy}z = -b\Phi(\bar{x}, \bar{y})\, b^{-1}z,$$

where b is regarded as a map $R^d \to M_m$.

(a) then follows from the fact that Φ is a 2-form, and hence is alternating, while (b) follows since Φ is $\mathfrak{o}(d)$-valued, $\mathfrak{o}(d)$ consisting of skew-symmetric transformations of R^d.

In order to prove (c), we first notice that for $a, b, c \in R^d$,

$$H[[E(a), E(b)], E(c)] = -H[\lambda\Phi(E(a), E(b)), E(c)] \qquad (6.1.4)$$

$$= -E(\Phi(E(a), E(b)) c) \qquad (6.2.1).$$

Choosing $\bar{x}, \bar{y}$ above in a particular way, we have

$$R_{xy}z = -b\Phi(E(b^{-1}(x))(b), E(b^{-1}(y))(b)) \, b^{-1}(z),$$

so that

$$E(b^{-1}R_{xy}z)(b) = -E(\Phi(E(b^{-1}(x))(b), E(b^{-1}(y))(b)) \, b^{-1} z)(b)$$

$$= H[[E(b^{-1}(x)), E(b^{-1}(y))], E(b^{-1}(z))](b).$$

But then the Jacobi identity gives

$$E(b^{-1}(R_{xy}z + R_{zx}y + R_{yz}x))(b) = 0,$$

which proves (c), since E and b are one-to-one.

(d) follows from (a), (b), (c) by taking inner products of equation (c) with w, then cyclicly permuting x, y, z, w. The four equations thus obtained are then added, and proper use of (a) and (b) will yield (d). Details are left to the reader.

Corollary 1. $K(P)$ is well defined.

Proof.   For $x, y \in M_m$, let $K(x, y) = \langle R_{xy}x, y \rangle / A(x, y)^2$. We point out that

(i) $K(x, y) = K(y, x),$

(ii) $K(ax, by) = K(x, y),$ if $ab \neq 0,$

(iii) $K(x + cy, y) = K(x, y).$

It then follows that if $x' = ax + by$, $y' = cx + dy$, $ad - bc \neq 0$, then $K(x', y') = K(x, y)$, since it is well known that the transformation from (x, y) to (x', y') can be obtained by a sequence of the types indicated in (i), (ii), and (iii).

Corollary 2. The $\langle R_{xy}x, y \rangle$ determine the curvature transformations.

Proof. More precisely, $\langle R_{xy}z, w \rangle$ is the only 4-linear function satisfying the properties of lemma 1 which restricts to $\langle R_{xy}x, y \rangle$. So we assume that we have two 4-linear functions, f and f', on M_m which satisfy the conditions corresponding to (a)-(d) and such that $f(x, y, x, y) = f'(x, y, x, y)$, all $x, y \in M_m$. Letting $g = f - f'$, we see that g satisfies these same conditions corresponding to (a)-(d). Replacing x by $x + z$ in $g(x, y, x, y) = 0$ we get

$$g(x, y, x, y) + g(z, y, x, y) + g(x, y, z, y) + g(z, y, z, y) = 0$$

and hence,

$$g(x, y, z, y) + g(z, y, x, y) = 0.$$

By (d)

$$g(x, y, z, y) = 0.$$

Replacing y by $y + w$ and following the same procedure gives

$$g(x, w, z, y) + g(x, y, z, w) = 0.$$

By (d), then (a), we get

$$g(x, y, z, w) = g(y, z, x, w),$$

so $g(., ., ., w)$ is invariant under cyclic permutation of the three entries. But the sum over such permutations is 0 by (c), so $g = 0$. QED

Remarks. (1) Sometimes it is more convenient to deal with curvature instead of curvature transformations and this corollary assures us that this will not lose information.

(2) If M has two Riemannian structures such that at a single point the inner product and curvature are the same, then the curvature transformations are the same.

(3) It is not correct to say that the curvature determines the curvature transformation, for two different Riemannian structures, with different curvature transformations, may give rise to the same curvature. For example, let $f: S^2 \rightarrow S^2$ be any diffeomorphism of the Riemannian 2-sphere. Viewing f as an isometry gives two Riemannian structures on S^2 with different curvature transformations but the same (constant) curvature.

Problem 2. Let $q(x, y) = f(x, y, x, y)$, $x, y \in M_m$. Establish the following explicit formula for f in terms of q:

$$6f(x, y, z, w) = q(x + z, y + w) - q(x + w, y + z)$$
$$+ q(x, y + z) - q(x, y + w) - q(y, x + z) + q(y, x + w)$$
$$- q(z, y + w) + q(z, x + w) - q(w, x + z) + q(w, y + z)$$
$$+ q(x, w) - q(x, z) + q(y, z) - q(y, w).$$

Problem 3. Use the following outline to prove *Schur's theorem* [*17*]: If K is constant on every fibre of $G_{d,2}(M)$, then K is constant on $G_{d,2}(M)$, for $d > 2$.

(a) This hypothesis is equivalent to: for every $x, y \in R^d$, $\langle \Phi(E(x), E(y)) x, y \rangle$ is constant on fibres of $F(M)$.

(b) Since the functions depending on x, y in (a) determine the functions $\langle \Phi(E(x), E(y)) z, w \rangle$, the hypothesis is equivalent to

$$\Phi(E(x), E(y)) \quad \text{is constant on fibres of } F(M).$$

(c) If $F\Phi(E(x), E(y)) = 0$ for every vertical F, $x, y \in R^d$, then $E(z) \Phi(E(x), E(y)) = 0$ for every $x, y, z \in R^d$, and hence $\Phi(E(x), E(y))$ is constant on $F(M)$, K constant on $G_{d,2}(M)$.

[*Hint:* Use the Bianchi identity

$$D\Phi(E(x), E(y), E(z)) = E(x) \Phi(E(y), E(z)) + E(y) \Phi(E(z), E(x))$$
$$+ E(z) \Phi(E(x), E(y)) = 0,$$

and the fact that

$$[\bar{A}, E(x)] = \bar{A}E(x) - E(x) \bar{A} = E(Ax) \qquad \text{for} \qquad A \in \mathfrak{o}(d),$$

so $E(x) \bar{A} + E(Ax) = \bar{A}E(x).$]

9.2 Computation of the Riemannian Curvature

We indicate briefly how the Riemannian curvature can be computed in terms of the metric coefficients g_{ij}. In particular, we show the connection between the curvature transformation and the metric.

By 6.4.3, if X, Y are vector fields, then

$$R_{XY} = \nabla_{[X,Y]} - [\nabla_X, \nabla_Y],$$

where ∇_X is the covariant derivative in the direction of X. However, by [66, p. 77] if X, Y, Z are vector fields,

$$2 \langle \nabla_X Y, Z \rangle = X \langle Y, Z \rangle + Y \langle X, Z \rangle - Z \langle X, Y \rangle$$
$$+ \langle [X, Y], Z \rangle + \langle [Z, X], Y \rangle + \langle X, [Z, Y] \rangle.$$

These two formulas give the desired connection.

Problem 4. The above formula depends on the following facts:

(1) Torsion zero if and only if $[X, Y] = \nabla_X Y - \nabla_Y X$, X, Y any C^∞ vector fields.

(2) Parallel translation preserves the inner product if and only if $X \langle Y, Z \rangle = \langle \nabla_X Y, Z \rangle + \langle Y, \nabla_X Z \rangle$. Prove these statements and the formula. Derive an explicit formula for $K(D_{x_i}, D_{x_j})$ in terms of the g_{ij}.

Problem 5. Use this formula to obtain an alternate proof for problem 7.18.

9.3 Continuity of the Riemannian Curvature

K is not a function on M, the Riemannian manifold, but it is a function on the Grassmann bundle of 2-planes of M [3.3(5)], and in fact a continuous function. From this it will follow that the curvature on a compact subset of M is bounded.

Let $G_{d,2}$ be the Grassmann manifold of plane sections (two-dimensional subspaces) of R^d. (See problem 7.30.) So

$$G_{d,2} = O(d)/O(2) \times O'(d-2).$$

We denote by $G_{d,2}(M)$ the bundle with fibre $G_{d,2}$ associated to the frame bundle $F(M)$, where M is a Riemannian manifold. Thus $G_{d,2}(M) = F(M) \times _{O(d)} G_{d,2}$. If $m \in M$, we write $G_{d,2}(m)$ for the fibre of $G_{d,2}(M)$ over m. If $b \in F(M)$ such that $\pi(b) = m$, then $b : G_{d,2} \xrightarrow{\approx} G_{d,2}(m)$ by $: P \to \{(b, P)\} = (b, P) O(d)$, the equivalence class of (b, P) in $G_{d,2}(M)$. But we know that $b : R^d \xrightarrow{\approx} M_m$, so $b : G_{d,2} \xrightarrow{\approx}$ {plane sections at m}, and the resulting identification of $G_{d,2}(m)$ with {plane sections at m} is independent of b. Hence, the Riemannian curvature K can be viewed as a real-valued function defined on $G_{d,2}(M)$. (We are here using the notation $F(M) \times _{O(d)} G_{d,2}$ for the space $(F(M) \times G_{d,2})/O(d)$ of 3.3.)

Proposition 1.   The function $K : G_{d,2}(M) \to R$ is C^∞, and hence, in particular, continuous.

Proof.   Consider the diagram

$$
\begin{array}{c}
F(M) \times O(d) \\
\downarrow p \\
F(M) \times G_{d,2} \\
\downarrow q \\
F(M) \times {}_{O(d)}G_{d,2} \xrightarrow{\ K\ } R
\end{array}
\qquad
\begin{array}{c}
\searrow \\
K \circ q \circ p
\end{array}
$$

p, q identification maps

It is only necessary to show that $K \circ q \circ p$ is C^∞. To define the map p we must first choose an element, say P_0, of $G_{d,2}$. Then $p(b, g) = (b, gP_0)$. Hence, $K \circ q \circ p(b, g) = K(b(gP_0))$, remembering that $b : G_{d,2} \to \{\text{plane sections at } m\}$. Let P_0 be spanned by orthonormal vectors $x, y \in R^d$. Then

$$
K(bgP_0) = \langle R_{b(gx)b(gy)}\, b(gx), b(gy) \rangle
$$
$$
= - \langle b\Phi(E(gx)(b), E(gy)(b))\, gx, gy \rangle,
$$

which is clearly C^∞ in b and g. QED

Since $G_{d,2}$ is compact, we have the:

Corollary.   If $C \subset M$ is compact, then there exist $H, L \in R$ such that for any plane section P at any point $m \in C$, $H \leqslant K(P) \leqslant L$.

Remark.   The curvature of a Riemannian manifold clearly depends on the particular Riemannian structure which the manifold is given. Thus, the flat torus has zero curvature everywhere, while the imbedded torus (doughnut) has points of both positive and negative curvature. However, a given manifold cannot admit arbitrary curvature. For example, it will be proved that a simply connected compact manifold cannot have everywhere nonpositive curvature (see corollary 2, theorem 4), while a noncompact complete manifold cannot have positive curvature bounded away from zero (Chapter 11). Moreover, there is a relationship between the curvature and the topological invariants of the manifold given by the Gauss-Bonnet theorem, which we do not study [*19, 22*]. In general, though, not much is known about this problem [*11, 12*].

Problem 6. A complete Riemannian manifold is *locally symmetric* if the curvature of a plane section is invariant under parallel translation of the plane section along geodesics. Show that this is equivalent to the vanishing of the covariant derivatives of the curvature transformation.

Problem 7. Let M be a Riemannian symmetric space. Show that the symmetry f_m of M_m carries a tangent into the negative of its parallel translate along the geodesic through m. Hence, show that M is locally symmetric. Conversely, it follows from a monodromy argument that a locally symmetric simply connected manifold is Riemannian symmetric.

Let M be a Riemannian symmetric space as above. From previous results (7.4.14) we know that M is a homogeneous symmetric space, that is, $M = G/H$, where G admits an involution f. Further, let $0 = eH \in M$, $\mathfrak{h} = \{X \in \mathfrak{g} \mid df(X) = X\}$, $\mathfrak{m} = \{X \in \mathfrak{g} \mid df(X) = -X\}$, then $\mathfrak{h}$ is the Lie algebra of H.

Problem 8. Prove the relations $[\mathfrak{h}, \mathfrak{h}] \subset \mathfrak{h}$, $[\mathfrak{h}, \mathfrak{m}] \subset \mathfrak{m}$, $[\mathfrak{m}, \mathfrak{m}] \subset \mathfrak{h}$.

Problem 9. Prove that for $X \in \mathfrak{m}$, $e^{tX} \cdot 0$ is a geodesic in M. [*Hint*: Show that if σ is a geodesic, the curve $\beta(t) = f_0 f_{\sigma(t)}$ in G satisfies

   (1) β is a one-parameter subgroup (of *transvections*)

   (2) β corresponds to an element of $\mathfrak{m}$.]

Show that there is hence an isomorphism between $\mathfrak{m}$ and M_0. Further, if the inner product $\langle \, , \, \rangle$ on M_0 is pulled back via this isomorphism to an inner product $(\, , \,)$ on $\mathfrak{m}$, then show that it satisfies

$$([X, Y], Z) + (Y, [X, Z]) = 0,$$

where $X \in \mathfrak{h}$, $Y, Z \in \mathfrak{m}$. (A form satisfying this relation is called *invariant* with respect to $\mathfrak{h}$. Cf. problem 7.23.)

Problem 10. Show that H is compact, using that it may be viewed as a closed subset of an orthogonal group. Hence, show that the form $(\, , \,)$ may be extended to an inner product $(\, , \,)$ on all of $\mathfrak{g}$ (see 7.4.11) which is invariant under df and with respect to $\mathfrak{h}$.

The Killing form $k(\, , \,)$ on $\mathfrak{g}$ is also invariant under df, since $\mathrm{ad}(df\, X) \circ \mathrm{ad}(df\, Y) = df \circ \mathrm{ad}\, X \circ \mathrm{ad}\, Y \circ df^{-1}$ (problem 7.23). Hence, show that $\mathfrak{m}$ and $\mathfrak{h}$ are orthogonal with respect to each of these forms.

Problem 11. M is said to be an *irreducible* symmetric space if ad $\mathfrak{h}\mid_\mathfrak{m}$ is real irreducible. In general, there is a linear transformation S_k of $\mathfrak{m}$ such that for $Y, Z \in \mathfrak{m}$,

$$k(Y, Z) = (S_k Y, Z).$$

Show that $\mathfrak{m}$ decomposes into the characteristic subspaces of S_k and that these subspaces are invariant under ad $\mathfrak{h}$. In particular, if M is irreducible there is a real number λ such that on $\mathfrak{m}$

$$k(\,,\,) = \lambda(\,,\,).$$

From problem 7.8 it follows that only the identity in H acts trivially on $\mathfrak{m}$, and, hence, if $X \in \mathfrak{h}$, $X \neq 0$, then ad $X\mid_\mathfrak{m} \neq 0$. Using this and the fact that ad X is skew-symmetric with respect to $(\,,\,)$ on $\mathfrak{g}$, show that $k(\,,\,)$ is negative definite on $\mathfrak{h}$. When M is irreducible with $\lambda \neq 0$ this shows that $k(\,,\,)$ is nondegenerate on $\mathfrak{g}$, from which it easily follows that $\mathfrak{g}$ has no proper Abelian ideals, that is, $\mathfrak{g}$ is *semisimple*. It can be shown that if $k(\,,\,)$ is negative definite, or $\lambda < 0$ in this case, then G is compact (see [33], p. 122).

Problem 12.   Pick $f \in F(M)$ such that $\pi(f) = 0$, and consider the corresponding imbedding of G as a closed submanifold of $F(M)$ (see problem 6.21). Show that G is a subbundle of $F(M)$ with group H, and hence that H may be viewed as a subgroup of $O(d)$.

Problem 13. G acts on $F(M)$ and so it makes sense to speak of invariant vector fields on $F(M)$. Show that basic and fundamental vector fields are invariant.

Problem 14.   Let $X \in \mathfrak{m}$, $Y \in \mathfrak{h}$. Show that X is the restriction to G of a basic vector field, and hence that the Riemannian connexion on $F(M)$ reduces to a connexion on G. This proves that the holonomy group of M with respect to f is contained in the isotropy group H. The converse is also true [18]. Show that Y is the restriction to G of a fundamental vector field, and in fact $Y = \lambda Y$, where $\mathfrak{h}$ is viewed as a subalgebra of $\mathfrak{o}(d)$. If $X = E(x)\mid_G$, $x \in R^d$, show that

$$\mathrm{ad}\ Y(X) = E(Yx)\mid_G.$$

Problem 15.   Let $X, Y, Z \in \mathfrak{m} \approx M_0$. Show that the curvature transformation is given by

$$R_{XY}Z = [[X, Y], Z].$$

Use this to derive a formula for the curvature of a plane section of M_0 in terms of the Lie algebra structure of $\mathfrak{g}$ and the inner product $(,)$. This then serves to determine the curvature everywhere on M.

Problem 16. Assume that M is irreducible. Show that if $\lambda = 0$, then curvature is 0; otherwise, curvature is nonnegative or nonpositive according as λ is negative or positive, respectively.

Problem 17. Calculate the curvature (constant) of the d-dimensional Riemannian sphere of radius r.

Examples. *Grassmann Manifolds*

Real case. If A is a matrix with real entries we let A^* denote the transpose of A. The Grassmann manifold $G_{d+e,d}$ of d-planes in R^{d+e} may be realized as the homogeneous space $O(d + e)/O(d) \times O(e)$. When split into corresponding sized blocks, the Lie algebra of $O(d + e)$ may be considered to be the matrices of the form $\left(\begin{smallmatrix} A & C \\ -C^* & B \end{smallmatrix}\right)$, where $A^* = -A$, $B^* = -B$. It is easily checked that $df\left(\begin{smallmatrix} A & C \\ -C^* & B \end{smallmatrix}\right) = \left(\begin{smallmatrix} A & -C \\ C^* & B \end{smallmatrix}\right)$ defines a Lie algebra involution with fixed algebra $\mathfrak{o}(d) + \mathfrak{o}(e)$ which generates an involution of $O(d + e)$ with fixed group $O(d) \times O(e)$.

If we define (X, X) to be the sum of the squares of entries of X, in this case $X \in \mathfrak{o}(d + e)$, $(X, X) = -\text{tr } X^2$, then $(,)$ is a positive definite form on $\mathfrak{o}(d + e)$ which is invariant under $\mathfrak{o}(d + e)$ and df. Thus $G_{d+e,d}$ becomes a Riemannian symmetric space.

The oriented Grassmann manifold

$$G'_{d+e,d} = SO(d + e)/SO(d) \times SO(e)$$

is a twofold covering of $G_{d+e,d}$, so it is also a Riemannian symmetric space.

Problem 18. Verify the unproved statements above. Find an explicit matrix formula for the involution f. Show that $G'_{d+e,d}$ has only one nontrivial isometry which corresponds to the identity on $G_{d+e,d}$. What are the groups of isometries of $G_{d+e,d}$ and $G'_{d+e,d}$? What are the isotropy groups of a point?

Complex and quaternion cases. We proceed as in the real case except that we let A^* be the transpose conjugate of A. The same formula for df is an involution of the real Lie algebras, and extends to an involution of the unitary or symplectic group to give the representations

$$H_{d+e,d} = \text{subspaces of dimension } d \text{ in } C^{d+e}$$
$$= U(d+e)/U(d) \times U(e).$$

$$K_{d+e,d} = \text{subspaces of dimension } d \text{ in } Q^{d+e}$$
$$= Sp(d+e)/Sp(d) \times Sp(e).$$

(For the definition of the symplectic group $Sp(d)$, see [25; p. 20].)

The sum of norms of entries is again a positive definite quadratic form on the Lie algebra which is invariant under the whole Lie algebra ($\mathfrak{u}(d+e)$ or $\mathfrak{sp}(d+e)$) and df. Thus each of these homogeneous spaces becomes a Riemannian symmetric space.

The tangent space at the basic point in each of these symmetric spaces may be identified with matrices of the form $\begin{pmatrix} 0 & C \\ -C^* & 0 \end{pmatrix}$, and then with C itself, where the entries of C are in the appropriate field. As such it has the structure of a complex or quaternion vector space. In the complex case this structure is invariant under the action of $\text{ad}(\mathfrak{u}(d) + \mathfrak{u}(e))$, so that it may be induced on every other tangent space invariantly by using the action of $U(d+e)$; thus $H_{d+e,d}$ has the structure of an almost complex manifold. (It is true that "almost" may be dropped.) We call those plane sections consisting of complex multiples of a single vector *holomorphic*. It is not true that the quaternion structure is invariant under $\text{ad}(\mathfrak{sp}(d) + \mathfrak{sp}(e))$, so that the quaternion structure cannot be induced invariantly on any tangent space of $K_{d+e,d}$.

Problem 19. Show that the symmetric spaces $G_{d+e,d}$, $H_{d+e,d}$, $K_{d+e,d}$ are irreducible with negative λ (see problem 11). Furthermore, they have nonnegative curvature, but not all positive curvature unless d or $e = 1$.

Problem 20. When $d = 1$ the Grassmann manifolds become projective spaces or spheres; show that the curvature may be described as follows, using problem 15:

Real field: All sections have the same curvature.

Complex field: If $X \perp Y$, X, $Y \in \mathfrak{m}$ and θ is the angle between the holomorphic sections of X and Y, then $K(X, Y) = 1 + 3 \cos^2 \theta$. In particular, only the holomorphic sections attain the maximal curvature of 4, and the so-called *holomorphic curvature* is constant.

Quaternion field: If $X \perp Y$, X, $Y \in \mathfrak{m}$, and θ is the angle between the 4-dimensional subspaces XQ and YQ, then $K(X, Y) = 1 + 3 \cos^2 \theta$.

Opposite spaces.   The Lie algebra of matrices of the form $\left(\begin{smallmatrix} A & C \\ C* & B \end{smallmatrix}\right)$, where $A^* = -A$, $B^* = -B$, with entries in one of the above-mentioned fields, has an involution given by

$$df \begin{pmatrix} A & C \\ C* & B \end{pmatrix} = \begin{pmatrix} A & -C \\ -C* & B \end{pmatrix}.$$

This involution extends to an involution of the connected subgroup G of $Gl(d + e, F)$ corresponding to the Lie algebra, and the fixed subgroup H may be divided by to give a homogeneous space M. Those matrices $\{X\}$ with $A = B = 0$ may be identified with the tangents to M at $0 = eH$, and the sum of norms quadratic form, in this case tr X^2, may be used to give M the structure of a Riemannian symmetric space. In this case the inner product on $\mathfrak{m}$ is invariant under only ad $\mathfrak{h}$, not under all of ad $\mathfrak{g}$ as with the Grassmann spaces.

Problem 21.   Show that these symmetric spaces are irreducible with positive λ. Moreover, the map which takes $\left(\begin{smallmatrix} 0 & C \\ -C* & 0 \end{smallmatrix}\right)$ into $\left(\begin{smallmatrix} 0 & C \\ C* & 0 \end{smallmatrix}\right)$ gives an isometry from the tangent space of the corresponding Grassmann space such that the curvature of corresponding plane sections under this isometry is the same in magnitude but opposite in sign.

In the case of the real field and $d = 1$ we obtain in this way *hyperbolic e-space*, R^e provided with a metric of constant negative curvature. Show that the map exp : $R^e \to M$ given by

$$\exp(X) = \exp \begin{pmatrix} 0 & X \\ X* & 0 \end{pmatrix} \cdot H \text{ is one-to-one.}$$

9.4 Rectangles and Jacobi Fields

Most of the theorems of this chapter as well as those of Chapter 11 are connected with the behavior of nearby geodesics or the infinitesimal analog thereof. Generally it is sufficient to consider a one-parameter family of such geodesics, and hence the natural setting for their study is a rectangle (8.1).

Let Q be a C^∞ rectangle in a Riemannian manifold M. We use the notation of 8.1 and the proof of therem 8.1.

Lemma 2. If the longitudinal curves of Q are geodesics, then we have the following formulas:

(a) $\qquad D_1\omega^Q(D_2) - D_2\omega^Q(D_1) = \phi^Q(D_2)\,\omega^Q(D_1),$

(b) $\qquad D_1\phi^Q(D_2) = \Phi^Q(D_1, D_2),$

(c) $\qquad D_1{}^2\omega^Q(D_2) = \Phi^Q(D_1, D_2)\,\omega^Q(D_1).$

(c) is a version of the Jacobi equation. [Φ^Q was not used in 8.1, but is, of course, $\bar{Q}^*\Phi$, where $\bar{Q}$ is the canonical lifting of Q to $f \in F(M)$.]

Proof. (a) has already been proved and used in theorem 8.1.
The second structural equation, theorem 6.4, gives

$$d\phi^Q = -\tfrac{1}{2}[\phi^Q, \phi^Q] + \Phi^Q.$$

Applying this to D_1, D_2 gives:

$$D_1\phi^Q(D_2) - D_2\phi^Q(D_1) - \phi^Q([D_1, D_2])$$
$$= -\tfrac{1}{2}([\phi^Q(D_1), \phi^Q(D_2)] - [\phi^Q(D_2), \phi^Q(D_1)]) + \Phi^Q(D_1, D_2).$$

But $\phi^Q(D_1) = 0$ and $[D_1, D_2] = 0$, so we have

$$D_1\phi^Q(D_2) = \Phi^Q(D_1, D_2),$$

which is (b).
Now apply D_1 to both sides of (a),

$$D_1{}^2\omega^Q(D_2) - D_1D_2\omega^Q(D_1) = (D_1\phi^Q(D_2))\,\omega^Q(D_2) + \phi^Q(D_2)(D_1\omega^Q(D_1)).$$

But $D_1\omega^Q(D_1) = 0$, and so also $D_1D_2\omega^Q(D_1) = D_2D_1\omega^Q(D_1) = 0$ since $[D_1, D_2] = 0$. Hence this becomes

$$D_1{}^2\omega^Q(D_2) = (D_1\phi^Q(D_2))\,\omega^Q(D_1). \quad \#$$

(c) now follows by substituting (b) into $\#$. QED

Jacobi fields provide the connection between the behavior of nearby curves and curvature. They are certain vector fields defined along a geodesic. Let σ be a geodesic and let V be a vector field along σ. [Actually V is a curve in $T(M)$ over σ.]
V is a *Jacobi field* if

$$\nabla_{\sigma_*}(\nabla_{\sigma_*}(V)) = -R_{\sigma_* V}\sigma_*.$$

Since V is defined only on σ, we shall often write V' for $\nabla_{\sigma_*}V$. So the Jacobi equation becomes $V'' = R_{V\sigma_*}\sigma_*$.

We shall reduce this to the classical equation.

Define functions R_{ijkl} on $F(M)$ by: $R_{ijkl} = \Phi_{ij}(E_k, E_l)$. Choose $b = (\sigma(0); e_1, ..., e_d) \in F(M)$ such that $e_d = \sigma_*(0)$. Let $\bar\sigma(t) = (\sigma(t); e_1(t), ..., e_d(t))$, $\bar{V}$ and $\bar{e}_d$ be the horizontal lifts of σ, V, and e_d respectively, and write $V(t) = \sum v_i(t) e_i(t)$. Then

$$R_{V(t)\sigma_*(t)}\sigma_*(t) = R_{V(t)e_d(t)}e_d(t)$$

$$= -\sum \Phi_{id}(\bar{V}(t), \bar{e}_d(t)) \, e_i(t) \qquad (6.1.5)$$

$$= -\sum \Phi_{id}\left(\sum v_k(t) E_k(\bar\sigma(t)), E_d(\bar\sigma(t))\right) e_i(t)$$

$$= -\sum v_k(t) R_{idkd}(\bar\sigma(t)) \, e_i(t).$$

On the other hand, $V''(t) = \sum v_i''(t) e_i(t)$, so the Jacobi equation becomes $v_i''(t) = -\sum_k v_k(t) R_{idkd}(\bar\sigma(t))$.

In particular $v_d''(t) = 0$, so the component of V in the direction of σ_* is a linear function of t. Thus if V is perpendicular to σ_* at two points, it is perpendicular everywhere. In any case, the behavior of the other v_i does not depend on v_d since $R_{iddd} = 0$.

In the two-dimensional case we have $v_1'' = -R_{1212}(\bar\sigma(t)) \, v_1(t)$, or $v'' + Kv = 0$, where K is the Gaussian curvature.

Since the Jacobi equation is a linear second order equation we immediately obtain from the theory of differential equations:

Proposition 2. The Jacobi fields along σ form a linear space of dimension $2d$ over R. For every $x, y \in M_{\sigma(0)}$ there is a unique Jacobi field V such that $V(0) = x$, and $V'(0) = y$. The Jacobi fields which vanish at $\sigma(0)$ form a linear subspace of dimension d whose values for sufficiently small t consist of all of $M_{\sigma(t)}$. If t is sufficiently small, for every $x \in M_{\sigma(0)}$ and $y \in M_{\sigma(t)}$ there is a unique Jacobi field V such that $V(0) = x$ and $V(t) = y$.

The following theorem characterizes Jacobi fields in a more geometric manner.

Theorem 1. A vector field V along a geodesic σ is a Jacobi field if and only if there is a rectangle Q having base curve σ, all longitudinal curves geodesics, and V its associated vector field.

Proof. If Q is such a rectangle, then the fact that V is a Jacobi field is really a corollary to lemma 2, for (c) is the Jacobi equation, only formulated in R^2. In fact, $V(t) = dQ(D_2(t, c))$, $\sigma_*(t) = dQ(D_1(t, c))$, so regarding $\bar{\sigma}(t)$ as a map : $R^d \rightarrow M_{\sigma(t)}$ we have

$$R_{V\sigma_*}\sigma_* = -\bar{\sigma}\Phi(\bar{V}, \bar{\sigma}_*)\bar{\sigma}^{-1}(\sigma_*) \qquad (6.1.5)$$

$$= -\bar{\sigma}\Phi^Q(D_2, D_1)\,\omega^Q(D_1) \circ j_c \qquad [j_c(t) = (t, c)]$$

$$= \bar{\sigma}D_1{}^2\omega^Q(D_2) \circ j_c \qquad\qquad [(c), \text{lemma } 2]$$

$$= V''.$$

On the other hand, if V is a Jacobi field, then by proposition 2 it will be sufficient to find a rectangle with longitudinal curves geodesics and associated vector field (along σ) W such that $W(0) = V(0)$ and $W'(0) = V'(0)$, for then W will be a Jacobi field by what we have just proved and so will agree with V everywhere by uniqueness.

Let γ be a curve such that $\gamma_*(0) = V(0)$, and let $\bar{\gamma}$ be the lift of γ starting at $f = (\sigma(0); f_1, ..., f_d) \in F(M)$, so $\bar{\gamma}(t) = (\gamma(t); f_1(t), ..., f_d(t))$. Let U be a curve above γ in $T(M)$, $U(t) = \Sigma h_i(t) f_i(t)$, such that $U(0) = \sigma_*(0)$ and $\Sigma h_i'(0) f_i = V'(0)$. Now we define the rectangle Q by

$$Q(s, t) = \exp_{\gamma(t)}sU(t).$$

It is clear that the longitudinal curves of Q are geodesics since for fixed t we just get the exponential map applied to a ray. Furthermore, since $U(0) = \sigma_*(0)$, the ray for $t = 0$ is the one belonging to σ, so σ is the base curve of Q. For $s = 0$ we get γ, so the associated vector field W satisfies $W(0) = \gamma_*(0) = V(0)$. Thus it only remains to show $W'(0) = V'(0)$.

Now Q satisfies the condition of lemma 2, so we have by (a) : $D_1\omega^Q(D_2) = D_2\omega^Q(D_1) + \phi^Q(D_2)\,\omega^Q(D_1)$. But $\bar{\gamma}$ is horizontal, so $\phi^Q(D_2(0, t)) = \phi(\bar{\gamma}_*(t)) = 0$. Also $\bar{\gamma}(t)\,\omega^Q(D_1(0, t))$ is the tangent to the longitudinal curve at height t, which from the definition of Q is $U(t)$. Thus for $t = 0$ we get

$$W'(0) = f(D_1\omega^Q(D_2)(0, 0))$$

$$= f(D_2\omega^Q(D_1)(0, 0))$$

$$= f(h'(0)) \qquad\qquad (h(t) = (h_1(t), ..., h_d(t)))$$

$$= V'(0). \qquad \text{QED}$$

We now apply this to the case where $V(0) = 0$ and Q is degenerate at $s = 0$, that is we take γ to be the constant curve. Thus in this case Q will be factored through $M_m (m = \sigma(0))$, that is, $Q = \exp_m \circ S$, where

S is a rectangle in M_m. Hence our Jacobi field will arise as $d \exp_m$ applied to a vector field along the ray ρ in M_m which goes into σ under $\exp_m$. We may taken U to be linear in t:

$$U(t) = \sigma_*(0) + tV'(0),$$

and this corresponds to

$$S(s, t) = s\sigma_*(0) + stV'(0).$$

A *linear homogeneous* vector field along a ray ρ in M_m is a curve X above ρ in $T(M_m)$ such that $X(0) = 0$ and $X'' = 0$ (differentiation is possible since M_m is a linear space).

For every $x \in (M_m)_0$ it is clear that there is a unique linear homogeneous vector field such that $X'(0) = x$. From the above definition of the rectangle S we thus have:

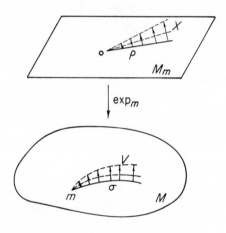

Corollary. If V is a Jacobi field along $\sigma = \exp_m \circ \rho$ which vanishes at m and X is the linear homogeneous vector field along ρ such that $V'(0) = d \exp_m X'(0)$, then

$$V = d \exp_m X.$$

In the remainder of this chapter we obtain results by making comparisons between the growth of X in the "flat" space M_m and the growth of V in M. This comparison will involve curvature, so according to theorem 1 we shall have a relation between curvature and the behavior of nearby geodesics. The metric on M_m is the one induced by the inner product, so M_m is isometric to R^d under, for example, $f \in F(M)$. We remark in passing that X is a Jacobi field along geodesic ρ in M_m.

We introduce the notation:

$$W = \rho_*, \qquad U = \sigma_*, \qquad K(V) = K(U,V) =$$

curvature of the section spanned by V and U when they are linearly independent.

Because the behavior of the U-component of V has been completely determined (it is a linear function of U) and because the part of V perpendicular to U is independent of the U-component, we assume for the remainder of the chapter that $\langle U, V \rangle = 0$. We assume also, for convenience, that $\| U \| = 1$.

Since $\| V \|'(0) = \| V' \|(0) = \| X \|'$ (see (a) in the proof of Theorem 2. We are here, of course, assuming that $V(0) = 0$), we can base our comparison of growth on the second derivative, which is given by the:

Lemma 3.

$$\| V \|'' = -K(V) \| V \| + \frac{A(V', V)^2}{\| V \|^3} .$$

[As in the definition of $K(P)$, $A(x, y)$ is the area of the parallelogram spanned by x and y.]

Notice that if $K(V)$ is negative, then $\| V \|''$ is positive.

Proof.

$$\| V \|' = (\tfrac{1}{2}) \langle V, V \rangle^{-1/2} (\langle V', V \rangle + \langle V, V' \rangle)$$
$$= \langle V', V \rangle \| V \|^{-1}$$

$$\| V \|'' = (\langle V'', V \rangle + \langle V', V' \rangle) \| V \|^{-1} - \langle V', V \rangle \| V \|^{-2} \langle V', V \rangle \| V \|^{-1}$$
$$= (\langle V', V' \rangle \langle V, V \rangle - \langle V', V \rangle^2) \| V \|^{-3} + \langle V'', V \rangle \| V \|^{-1}$$
$$= A(V', V)^2 \| V \|^{-3} - \langle R_{UV} U, V \rangle \| V \|^{-1} \quad (V \text{ is a Jacobi field})$$
$$= A(V', V)^2 \| V \|^{-3} - K(V) A(U, V)^2 \| V \|^{-1}.$$

But $A(U, V) = \| V \|$ by the assumptions $\langle U, V \rangle = 0$ and $\| U \| = 1$.

QED

Problem 22. Determine the Jacobi fields on R^d (see problem 7.9).

9.5 Theorems Involving Curvature

We use the notation of 9.4.

Theorem 2. Let ρ be a ray through 0 in M_m .

(i) If $K(V) \leqslant 0$ along ρ, then $\| V \| = \| d \exp_m X \| \geqslant \| X \|$ along ρ. Further, strict inequality is preserved in this implication.

(ii) If $K(V') > 0$ at 0, then $\| d \exp_m X \| < \| X \|$ at points of ρ near 0.

Proof. Let $g = \| V \| - \| X \|$. Then since $\| X \| (t) = t \| X' \|$ for $t \geqslant 0$, it is clear that g' exists on the positive real numbers whenever $\| V \| \neq 0$. Thus it will suffice to show

(a) g has right-hand derivative 0 at 0.

(b) If $K(V) \leqslant 0$, then $g'' \geqslant 0$, and the same implication for strict inequality.

(c) If $K(V')(0) > 0$, then $g'' < 0$ in a neighborhood of 0 in the positive reals.

For (a) we have: since $X(t) = tX'(t)$,

$$\| V \| (t) = \| d \exp_m tX'(t) \| = t \| d \exp_m X'(t) \|.$$

Dividing by t and taking the limit to $0 +$ we get

$$\| V \|'(0 +) = \| d \exp_m X'(0 +) \| = \| X'(0 +) \|$$

since $\exp_m$ is an isometry on $(M_m)_0$.

Since $\| X \|'' = 0$, (b) is immediate from lemma 3.

To show (c) we write the formula for $\| V \|''$ as

$$\| V \|'' = \left(-K(V) + \frac{A(V', V)^2}{\| V \|^4} \right) \| V \|.$$

Since $\lim K(V) = \lim K(V/t) = K(V'(0))$ by the continuity of K, it suffices to show that the limit of $A(V', V)^2 \langle V, V \rangle^{-2}$ is zero. To do this we let $Y = V/u$, so $Y = d \exp_m X'$, $V' = Y + uY'$, $V'' = 2Y' + uY'' = -R_{UV}U$ (u is the coordinate on R). By the fact that uY and $Y + uY'$ span a parallelogram of the same area as uY and uY' and because $A(x, y)^2$ is quadratic in each variable, we have

$$A(V', V)^2 \langle V, V \rangle^{-2} = A(uY, uY')^2 \langle uY, uY \rangle^{-2}$$
$$= A(Y, Y')^2 \langle Y, Y \rangle^{-2}.$$

Now the limit of Y at 0 is $d\,\exp_m X'(0) = V'(0) \neq 0$, whereas the limit of Y', from the equation for V'', is 0. Thus the limit of $A(Y, Y')^2 \langle Y, Y \rangle^{-2}$ is 0. QED

Corollary 1. If M has everywhere nonpositive curvature, then $\exp_m$ cannot decrease lengths of curves.

Proof. By (i), since we can take σ to be any geodesic from m and X any linear homogeneous vector field perpendicular to ρ, it follows that $\exp_m$ cannot decrease lengths of vectors perpendicular to ρ. But $\| d\,\exp_m U \| = \| U \|$, so $\exp_m$ cannot decrease the length of any vector.

Corollary 2. If $K(V) \leqslant 0$ for all Jacobi fields along σ, then geodesics from m near σ pull away from σ as compared to the corresponding rays in M_m.

$K \leqslant 0$ $K > 0$

Fig. 33.

If $K(V) > 0$ along σ near m, then geodesics from m near σ pull towards σ as compared to the corresponding rays in M_m.

Proof. We need only make precise the words. Let ρ, τ be rays in M_m through 0. Let S be a sphere about 0, and let γ be the shorter segment of the great circle in S intersecting ρ and τ. Then the geodesics $\exp_m \circ \rho$ and $\exp_m \circ \tau$ pull together if $|\gamma| \geqslant |\exp_m \circ \gamma|$ and pull apart if $|\gamma| \leqslant |\exp_m \circ \gamma|$. The corollary then follows from theorem 2.

Problem 23. Plane curvature can be used as follows to give a comparison between the lengths and areas of "circles" in M and those in R^2. Let $x, y \in M_m$, x, y unit orthogonal. Define a C^∞ rectangle Q by

$$Q(s, t) = \exp_m (s(x \cos t + y \sin t)),$$

$0 \leqslant s, 0 \leqslant t \leqslant 2\pi$. Let $K = K(x, y)$, $V = $ the transverse vector field of Q.

(a) Show that the length of V may be expressed as

$$\| V \| = s - \tfrac{1}{6} Ks^3 + s^4 h(s, t),$$

where h is C^∞ for $s > 0$, continuous at $s = 0$.

(b) Let $L(s)$ be the length of the "circle" Q_s. Show that $L(s) = 2\pi(s - Ks^3/6) + s^4 f(s)$, where f is a continuous function of s; hence

$$K = (3/\pi) \lim_{s \to 0} (2\pi s - L(s))/s^3.$$

(c) Define $A(s) = \int_0^s L(u) \, du$, the "area" of the "circle," and obtain the formula

$$K = (12/\pi) \lim_{s \to 0} (\pi s^2 - A(s))/s^4.$$

Remark. É. Cartan gives a generalization which tells the area of small spheres, volume of small 3-spheres, etc. [*17*, p. 252].

Problem 24. Let $M = S^d$ of radius r. Using problem 7.12 show that the "circles" of "radius s in S^{d}" are circles of radius $r \sin s/r$ in R^{d+1}. Hence, calculate $L(s)$ and prove that $K = 1/r^2$ for all plane sections.

Problem 25. Use the above to calculate explicitly the Jacobi fields on S^d as follows: let $\gamma(s)$ be a geodesic with s arc length. Let x, $y \in S^d_{\gamma(0)}$ be tangents perpendicular to $\gamma_*(0)$. Then identifying tangents to S^d with tangents to R^{d+1}, we may write

$$x = \sum a_i D_i(\gamma(0))$$

$$y = \sum b_i D_i(\gamma(0)).$$

Show that the formula for the Jacobi field X along γ with $X(0) = x$, $X'(0) = y$ is

$$X(s) = \cos s/r \sum a_i D_i(\gamma(s)) + \sin s/r \sum b_i D_i(\gamma(s)).$$

Let M, $'M$ be d-dimensional Riemannian manifolds, $m \in M$, $'m \in {}'M$, $f = (m, f_1, ..., f_d) \in F(M)$, $'f = ('m, {}'f_1, ..., {}'f_d) \in F('M)$, and define as in 8.1

$$\theta = \overline{\exp_f}^* \phi \qquad '\theta = \overline{\exp_{'f}}^{*'} \phi$$

$$\Theta = \overline{\exp_f}^* \Phi \qquad '\Theta = \overline{\exp_{'f}}^{*'} \Phi$$

$$\psi = \overline{\exp_f}^* \omega \qquad '\psi = \overline{\exp_{'f}}^{*'} \omega.$$

Let $X \in \mathfrak{o}(d)$; then $f \circ (\exp uX) \circ f^{-1}$ is a one-parameter group of rotations of M_m and thus induces a vector field $\bar{X}$ on M_m. Using $'f$ instead of f we likewise get a vector field $'\bar{X}$ on $'M_{'m}$. The vector field $\bar{X}$ is linear homogeneous along and perpendicular to every ray ρ in M_m; further $(\bar{X} \circ \rho)'(0) = d \exp_m^{-1}(fXf^{-1}(\rho(1)))$.

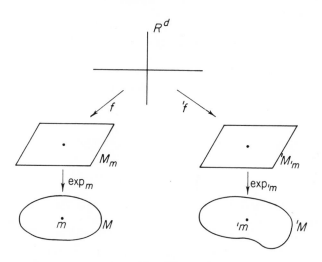

FIG. 34.

Let T be the radial vector field defined on all of M_m except 0 which satisfies $T \circ \rho = U$ whenever ρ and U are as before. Similarly we define $'T$ on $'M_{'m}$.

Theorem 3 (É. Cartan). If for every $x \in N$, a neighborhood of 0 in R^d, and for every $X \in \mathfrak{o}(d)$ we have

$$\Theta(T, \bar{X})(fx) = '\Theta('T, '\bar{X})('fx),$$

then there is a neighborhood U of m and an isometry $J : U \to 'M$ such that $J(m) = 'm$ and $dJ \circ f = 'f$.

Proof. Let L be the linear map $'ff^{-1} : M_m \to 'M_{'m}$. Let $\bar{U}$ be a neighborhood of 0 in M_m on which $\exp_m$ is a diffeomorphism and which is contained in $f(N)$. Then we set $U = \exp_m \bar{U}$ and define J by $J = \exp_{'m} \circ L \circ (\exp_m | \bar{U})^{-1}$. It is then clear that $J(m) = 'm$ and $dJ = L$ on M_m, so it only remains to show that J is an isometry.

It is clear that dJ preserves lengths of radial vectors, that is, $dJ(d\exp_m T) = d\exp_{'m}' T$. It therefore suffices to show that dJ preserves lengths of vectors normal to the radial ones. Now if t is such a tangent to U, then there is $p \in M_m$ and $X \in \mathfrak{o}(d)$ such that $d\exp_m \bar{X}(p) = t$, by the choice of U. Also, $dJ(t) = d\exp_{'m}' \bar{X}(Lp)$. Therefore, by formula (b), p. 146, $\| t \| = \| \psi(\bar{X})(p) \|$, $\| dJ(t) \| = \| {}'\psi({}'\bar{X})(Lp) \|$, and hence it will be sufficient to prove

$$\psi(\bar{X}) = {}'\psi({}'\bar{X}) \circ L \qquad \text{for every } X.$$

Now we make the observations:

(a) ψ and $f^{-1} \circ d\exp_m$ are the same on $(M_m)_0$, and similarly for ${}'\psi$.

(b) $\psi(T) = {}'\psi({}'T) \circ L$

(c) $\psi(\bar{X})(0) = {}'\psi({}'\bar{X})(L(0)) = 0$.

For every ray ρ in M_m,

(d) $(\psi(\bar{X}) \circ \rho)'(0) = \psi((\bar{X} \circ \rho)'(0)) = Xf^{-1}\rho(1)$
$\qquad = ({}'\psi({}'\bar{X}) \circ L \circ \rho)'(0),$

by (a) and the remark before the statement of the theorem describing $(\bar{X} \circ \rho)'(0)$.

By (c) and (d) $\psi(\bar{X})$ and ${}'\psi({}'\bar{X}) \circ L$ satisfy the same initial conditions along any ray. They will be the same if we can show that the second derivatives are the same. But rays are integral curves of T, so we must show that $T^2\psi(\bar{X})$ and $T^2({}'\psi({}'\bar{X}) \circ L)$ coincide. Deferring the proof that $T^2\psi(\bar{X}) = \Theta(T, \bar{X})\psi(T)$, we have

$$\begin{aligned}
T^2({}'\psi({}'\bar{X}) \circ L) &= ({}'T^{2'}\psi({}'\bar{X})) \circ L \\
&= ({}'\Theta({}'T, {}'\bar{X}){}'\psi({}'T)) \circ L \\
&= ({}'\Theta({}'T, {}'\bar{X}) \circ L)({}'\psi({}'T) \circ L) \\
&= \Theta(T, \bar{X})\psi(T) \qquad \text{by hypothesis and by (b)} \\
&= T^2\psi(\bar{X}).
\end{aligned}$$

The following lemma will complete the proof:

Lemma 4.   (a') $T\psi(\bar{X}) = \bar{X}\psi(T) + \theta(\bar{X})\psi(T),$
$\qquad\qquad$ (b') $T\theta(\bar{X}) = \Theta(T, \bar{X}),$
$\qquad\qquad$ (c') $T^2\psi(\bar{X}) = \Theta(T, \bar{X})\psi(T).$

Proof.   Except for an obvious substitution of symbols, the proof is the same as that of lemma 2. The only fact we need to establish is

$[T, \bar{X}] = 0$. This follows easily from the geometric interpretation of bracket, theorem 1.4, for the same result is obtained from either order of the operations

   moving a given distance along a ray (following an integral curve of T)

   rotating about the origin (following an integral curve of $\bar{X}$).

Corollary.   If M is a flat d-dimensional Riemannian manifold, that is, $K(P) = 0$ for all plane sections P on M, then M is locally isometric to R^d.

   In the following, we assume our manifolds are connected.

Theorem 4.   Let M be a complete Riemannian manifold, and let $m \in M$. Assume that $d \exp_m$ is everywhere nonsingular. Then $\exp_m$ is a covering map.

Proof.   First define a new Riemannian metric on M_m [that is, on $T(M_m)$] by pulling back the metric on M by means of $d \exp_m$. Since geodesics from $0 \in M_m$ are straight lines with linear parametrization it follows from theorem 8.5 that M_m is complete in this metric.

   Therefore, to prove the theorem it is sufficient to prove that if $F : N \to M$ is a local diffeomorphism with dF everywhere an isometry, N complete, then F is a covering map. For this we need to show that for any $m \in M$, there is a neighborhood U of m which is evenly covered by F, that is, $F^{-1}(U)$ is the disjoint union of sets U_i, each of which is diffeomorphic to U under F.

   Fix $m \in M$. Let O be a ball about m which is diffeomorphic to a ball O' of radius r about $0(m)$ in M_m. Let U be the ball of radius $r/2$ about m, let $F^{-1}(m) = \{u_i\}$, and let U_i be the ball of radius $r/2$ about u_i. To show that this is an even covering we must prove that

   (1) $F \mid_{u_i}$ is 1-1,
   (2) $F^{-1}(U)$ is the union of the U_i,
   (3) $u_i \neq u_j$ implies U_i and U_j are disjoint.

Proof of (1).   Let U' be the ball of radius $r/2$ about $0(m)$ in M_m, and let dF_{u_i} map N_{u_i} isometrically onto M_m. Then from the relation $\exp_m \circ dF_{u_i} = F \circ \exp_{u_i}$ we have that $U_i = \exp_{u_i} \circ dF_{u_i}^{-1}(U')$ and $F \mid U_i$ is 1-1, since dF_{u_i} is an isometry.

Proof of (2).   Let $u' \in F^{-1}(U)$. We want to show that there exists an i such that $u' \in U_i$. Let $m' = F(u')$, and let σ be the geodesic segment of length $k < r/2$ from m' to m. Locally, σ can be lifted to a geodesic $\bar{\sigma}$ from u'. Since N is complete, $\bar{\sigma}$ can be extended to length k, and since F is a local isometry, $F \circ \bar{\sigma}$ is a geodesic and hence is σ again. But then the end point of $\bar{\sigma}$, $\bar{n}$ say, goes into the end point of σ, that is, $F(\bar{n}) = m$. Hence, there exists an i such that $u_i = \bar{n}$, and so $u' \in U_i$. This proves (2).

Proof of (3). Assume that $u_i \neq u_j$ and $u' \in U_i \cap U_j$. Then

$$\rho(u', u_i) < r/2, \rho(u', u_j) < r/2 \quad \text{so} \quad \rho(u_i, u_j) < r.$$

But by reasoning similar to that of (1), F is 1-1 on a ball of radius r about u_i, which contradicts $F(u_i) = F(u_j)$. This proves (3) and the theorem.

Corollary 1. Let M be a complete simply connected d-dimensional flat Riemannian manifold. Then M is isometric to R^d.

Proof. By the corollary to theorem 3, $\exp_m : M_m \to M$ is a local isometry, for any $m \in M$, and hence by theorem 4 is a covering map, which is 1-1 since M is simply connected.

Corollary 2. (Hadamard-Cartan). Let M be a complete simply connected d-dimensional Riemannian manifold with nonpositive curvature at all plane sections. Then M is diffeomorphic to R^d.

Proof. This follows immediately from theorems 2 and 4.

Problem 26. Assume N, M are Riemannian manifolds, N complete, $\phi : N \to M$ regular and locally distance-nondecreasing. Show that ϕ is a covering map if $\dim N = \dim M$.

Problem 27. Let M and N be complete and have the same constant curvature. (Such manifolds are called *space forms*.)

    (a) Show that the curvature form may be written $\Phi = k\omega\omega^t$, where ω^t is the transpose of ω.

    (b) Show that M and N are locally isometric.

    (c) If M and N are simply connected and $k \leq 0$, then they are isometric (M and N connected).

    (d) If $k = a^2 > 0$, then the sphere of radius π/a in M_m is mapped to a point by $\exp_m$, and $\exp_m$ is regular within that sphere. Construct a Riemannian covering map from the sphere S^d of radius $1/a$ to M, so that if M is simply connected it is isometric to S^d.

CHAPTER 10

Immersions and the Second Fundamental Form

In this chapter we consider the immersion of a manifold in a Riemannian manifold and the resulting induced Riemannian connexion on the immersed manifold. The second fundamental form is defined and related to curvature and parallel translation, and Synge's theorem is proved. The problem of the existence of immersions is formulated, and the chapter concludes with a section on hypersurfaces [20, 38, 50, 83].

10.1 Definitions

Let N be an f-dimensional Riemannian manifold with metric $\langle\,,\,\rangle$ and M a d-dimensional manifold with $e = f - d > 0$.

An *immersion* of M into N is a C^∞ map $I : M \to N$ such that dI is one-to-one on every M_m. Recall that I is an imbedding if it is one-to-one (1.5).

The *induced Riemannian metric* (or *first fundamental form of the immersion*) is $\langle\,,\,\rangle' = \langle dI,\, dI\rangle$, which makes M a Riemannian manifold.

We assume from now that I is an immersion and that M has the induced Riemannian structure, and so also the connexion.

N and M being Riemannian manifolds, they have associated frame bundles $F(N)$ and $F(M)$. We consider two more bundles. First, let $F_M(N)$ be the principal bundle on M induced by I and $F(N)$. Thus,

$$F_M(N) = \{(m, b) \mid I(m) = \pi_N(b), \quad m \in M, \quad b \in F(N)\};$$

we shall suppress the $I(m)$ in $b = (I(m), e_1, ..., e_f)$ and write instead $(m, b) = (m, e_1, ..., e_f)$.

$O(f)$ acts on R^f and is the group of the bundles $F(N)$ and $F_M(N)$. Decompose R^f into $R^d + R^e$, and consider the following subgroups of $O(f)$:

$$O(d) = \{g \in O(f) \mid gR^d \subset R^d, \quad g \mid_{R^e} = \text{identity}\},$$
$$O(e) = \{g \in O(f) \mid g \mid_{R^d} = \text{identity}, \quad gR^e \subset R^e\}.$$

Let $\mathfrak{o}(f)$, $\mathfrak{o}(d)$, $\mathfrak{o}(e)$ be the corresponding Lie algebras, so that $\mathfrak{o}(f) = \mathfrak{o}(d) + \mathfrak{o}(e) + \mathfrak{k}$, where

$$\mathfrak{k} = \{X \in \mathfrak{o}(f) \mid XR^d \subset R^e \text{ and } XR^e \subset R^d\}.$$

Notice that $O(d) \times O(e) \subset O(f)$ is a subgroup, and

$$\text{Ad}\,(0(d) \times 0(e))\, \mathfrak{k} \subset \mathfrak{k}.$$

Let $\perp(M)$, $T(M)$ be the normal and tangent bundles of M, respectively, with fibres $M_m^\perp$ and M_m. Now define

$$F(N, M) = \{b \in F_M(N) \mid b(R^d) = d\,IM_m, \quad m = \pi(b)\} \qquad (3.3);$$

that is,

$$F(N, M) = \{(m, e_1, \ldots, e_f) \in F_M(N) \mid e_1, \ldots, e_d \in M_m, e_{d+1}, \ldots, e_f \in M_m^\perp\}.$$

$F(N, M)$ is called the bundle of *adapted frames* over M. It realizes a reduction of the group $O(f)$ of the bundle $F_M(N)$ to $O(d) \times O(e)$.

We now have the commutative diagram:

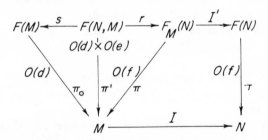

where s is the projection onto the first d tangents.

Problem 1. Let S be the vector space of symmetric $d \times d$ matrices over R. Then $O(d)$ acts on S by the restriction of the adjoint representation of $Gl(d, R)$: for $T \in O(d)$, $X \in S$, $\text{Ad}\,T(X) = TXT^{-1}$. Let I_r be the $r \times r$ identity matrix,

$$X_r = \begin{pmatrix} I_r & 0 \\ 0 & -I_{d-r} \end{pmatrix}.$$

(a) Let M be the orbit of X_r under $O(d)$. Show that the isotropy group which leaves X_r fixed is $O(r) \times O(d - r)$, so M is diffeomorphic to $G_{d,r}$.

(b) If we introduce the metric $\langle\ ,\ \rangle$ on S defined by $\langle X, X' \rangle = \operatorname{tr} XX'$, then S is isometric to flat R^D space, $D = d(d + 1)/2$, and $O(d)$ acts as isometries. The metric induced on M is thus invariant under $O(d)$, so M, with this metric, is a Riemannian symmetric space.

Problem 2. If we let $O(d)$ act on S as in problem 1, and $X \in S$ has n distinct characteristic values with multiplicities $d_1, ..., d_n$, show that the orbit is the flag manifold $Fl(d; d_1, ..., d_n)$. (See problem 7.25.) Establish the relation between the inner products on the blocks m_{ij} and the characteristic values of X.

10.2 The Connexions

We shall shortly wish to induce from a connexion on $F_N(M)$ one on the subbundle $F(N, M)$. In general, the horizontal distribution on $F_M(N)$ will not be tangent to $F(N, M)$, and hence it will have to be "projected" onto $F(N, M)$ in some sense. We describe a procedure for carrying this out, using the dual formulation for connexion.

Let (P, G, M) be a principal bundle over a manifold M and let ϕ be a connexion form $\phi : T(P) \to \mathfrak{g}$. Let (B, H, M) be a subbundle of (P, G, M), $i : B \subset P$, $H \subset G$, representing a reduction of G to H. Then $i^*\phi$ is a 1-form on B taking values in $\mathfrak{g}$, and hence is not in general a connexion form on B. However, if $\mathfrak{k}$ is a vector complement of $\mathfrak{h}$ in $\mathfrak{g}$ invariant under $\operatorname{Ad} H$, that is, $\mathfrak{g} = \mathfrak{k} + \mathfrak{h}$, $\operatorname{Ad} H(\mathfrak{k}) \subset \mathfrak{k}$, then the projection of $i^*\phi$ into $\mathfrak{h}$ under this decomposition of $\mathfrak{g}$ is a connexion form on B (see problem 5.5).

Problem 3. Verify that under the condition stated, the projection of $i^*\phi$ into $\mathfrak{h}$ is a connexion form.

Let ϕ be the Riemannian connexion form on $F(N)$, ω the solder form, and denote by the same symbols these forms pulled back to $F_M(N)$ by I', so ϕ is a connexion form on $F_M(N)$, and $d\omega = -\phi\omega$ is the first structural equation.

Replacing $\mathfrak{g}$, $\mathfrak{h}$, $\mathfrak{k}$ by $\mathfrak{o}(f)$, $\mathfrak{o}(d) + \mathfrak{o}(e)$, $\mathfrak{k}$ in the above, we see that the projection of $r^*\phi$ onto $\mathfrak{o}(d) + \mathfrak{o}(e)$ is a connexion form on $F(N, M)$, denoted by $\phi_d + \phi_e$, with the obvious meanings. This again gives a connexion form ψ on $F_M(N)$, by equivariance; and so we may consider the difference form $\phi - \psi$, which is horizontal and equivariant. Thus $\tau = r^*\phi - (\phi_d + \phi_e)$ is also horizontal and equivariant. τ is essentially the second fundamental form of the immersion, although the formal definition will be in terms of a related object.

We now use ϕ, to define a connexion on $F(M)$. Notice that ϕ_d satisfies

(1) $\phi_d(ds^{-1}(0)) = 0$

(2) $R_g^*\phi_d = \phi_d$, for $g \in I_d \cdot O(e)$.

From (1) and (2) it follows that there exists a unique form ϕ_0 on $F(M)$ such that

$$\phi_d = \phi_0 \circ ds.$$

It is easy to verify that ϕ_0 is a connexion form on $F(M)$.

We now study the structural equations of these various connexions. First notice that if $b \in F(N, M)$, then

$$d\pi'(F(N, M)_b) \subset M_{\pi'(b)}$$

and

$$b^{-1}(M_{\pi'(b)}) \subset R^d.$$

Therefore, if we let $\omega' = r^*\omega$,

$$\omega'(F(N, M)_b) = b^{-1}(d\pi'(F(N, M)_b)) \subset R^d,$$

where ω is the solder form on $F(N)$ pulled back to $F_M(N)$. Hence, $d\omega' = d(r^*\omega) = r^* \, d\omega$ also takes values in R^d, and $\phi_e\omega' = 0$. So the structural equation for ϕ on $F_M(N)$ gives

$$d\omega' = r^* \, d\omega = -r^*\phi\omega = -(\phi_d + \phi_e + \tau) \, \omega'$$
$$= - \phi_d\omega' - \tau\omega'.$$

Now, $d\omega'$ and $-\phi_d\omega'$ are R^d-valued, while $\tau\omega'$ is R^e-valued, since ω' is R^d-valued. Hence,

$$d\omega' = -\phi_d\omega'$$
$$\tau\omega' = 0.$$

In particular, the connexion $\phi_d + \phi_e$ has zero torsion. Further, since $\omega' = s^*\omega_0$, where ω_0 is the solder form of $F(M)$, the structural equation of ϕ_0 is

$$d\omega_0 = -\phi_0\omega_0,$$

so that ϕ_0 has zero torsion. Therefore, ϕ_0 is the unique Riemannian connexion on $F(M)$.

Problem 4.   Prove that $\omega' = s^*\omega_0$.

Problem 5.   If we define a Riemannian metric on $F(N)$ by letting the basic and fundamental vector fields, E_i , F_{ij} , be orthonormal, this induces a metric on $F_M(N)$.

(a) Show that the subspace $dI'(F(N, M)_b) \subset F_M(N)_b$ is orthogonal to the part of the vertical space which corresponds to $\mathfrak{k}$, that is, it is orthogonal to $\lambda \mathfrak{k}$.

(b) Show that the horizontal space of the ψ connexion at $b \in F_M(N)$ is the orthogonal projection of the horizontal space of the ϕ connexion onto $dI'(F(N, M)_b)$.

This result provides a more geometric description of the ψ connexion: it is the closest connexion to the ϕ connexion which parallel-translates adapted frames into adapted frames.

10.3 Curvature

We now employ the other structural equation of ϕ to relate the curvature of the induced structure on M to the curvature of N.

Note that

(1) $[\phi_d , \phi_e] = 0$

(2) $[\phi_d + \phi_e , \tau]$ is $\mathfrak{k}$-valued

(3) $[\phi_d , \phi_d]$ is $\mathfrak{o}(d)$-valued

(4) $[\phi_e , \phi_e]$ is $\mathfrak{o}(e)$-valued

(5) $[\tau, \tau]$ is $\mathfrak{o}(d) + \mathfrak{o}(e)$-valued, $[\tau, \tau] = [\tau, \tau]_d + [\tau, \tau]_e$.

These facts are all trivial.

Now let Φ be the curvature form of ϕ on $F_M(N)$, and write $\Phi' = r^*\Phi = \Phi_d + \Phi_e + \Gamma$, where Γ takes values in $\mathfrak{k}$. Then the structural equation

$$d\phi = -\tfrac{1}{2}[\phi, \phi] + \Phi$$

gives the equations

$$d\phi_d = -\tfrac{1}{2}[\phi_d , \phi_d] - \tfrac{1}{2}[\tau, \tau]_d + \Phi_d$$
$$d\phi_e = -\tfrac{1}{2}[\phi_e , \phi_e] - \tfrac{1}{2}[\tau, \tau]_e + \Phi_e$$
$$d\tau = -[\phi_d + \phi_e , \tau] + \Gamma.$$

In particular, if Φ_0 is the curvature form of ϕ_0 on $F(M)$, then

$$\Phi_0 \circ ds = \Phi_d - \tfrac{1}{2}[\tau, \tau]_d.$$

Thus, the curvatures are related by means of the difference form τ.

More precisely, let x, y be orthonormal tangents in M_m, $b \in F(N, M)$ with $\pi'(b) = m$, $\tilde{x}, \tilde{y}$ horizontal liftings of x, y to b, so that $\tilde{x} = ds(\tilde{x})$, $\tilde{y} = ds(\tilde{y})$ are ϕ_0 horizontal lifts of x, y to $s(b) \in F(M)$. Denote by $K_0(x, y)$ and $K(x, y)$ the curvatures of the section spanned by x and y in M and N, respectively.

Then

$$
\begin{aligned}
K_0(x, y) &= \langle R_{0xy}x, y \rangle \\
&= - \langle s(b)\, \Phi_0(\tilde{x}, \tilde{y})(s(b))^{-1}x, y \rangle \\
&= - \langle s(b)\, \Phi_0(ds(\tilde{x}), ds(\tilde{y}))(s(b))^{-1}x, y \rangle \\
&= - \langle b\Phi_d(\tilde{x}, \tilde{y})\, b^{-1}x, y \rangle + \tfrac{1}{2}\langle b[\tau, \tau](\tilde{x}, \tilde{y})\, b^{-1}x, y \rangle.
\end{aligned}
$$

But $(\Phi_e(\tilde{x}, \tilde{y}) + \Gamma(\tilde{x}, \tilde{y}))\, b^{-1}x \in R^e$, since $b^{-1}x \in R^d$. Therefore, since $\langle R^d, R^e \rangle = 0$,

$$\langle b\Phi_d(\tilde{x}, \tilde{y})\, b^{-1}x, y \rangle = \langle b\Phi'(\tilde{x}, \tilde{y})\, b^{-1}x, y \rangle = -K(x, y),$$

so

$$K_0(x, y) = K(x, y) + \tfrac{1}{2}\langle b[\tau, \tau](\tilde{x}, \tilde{y})\, b^{-1}x, y \rangle,$$

where

$$[\tau, \tau](\tilde{x}, \tilde{y}) = 2[\tau(\tilde{x}), \tau(\tilde{y})].$$

Thus,

$$\tfrac{1}{2}[\tau, \tau](\tilde{x}, \tilde{y})\, b^{-1}x = [\tau(\tilde{x}), \tau(\tilde{y})]\, \omega'(\tilde{x}).$$

This last term will be given geometrical significance via the second fundamental form.

10.4 The Second Fundamental Form

Let $z \in M_m^\perp$. The *second fundamental form of* z is a bilinear form H_z on M_m defined by

$$H_z(x, y) = - \langle z, b\tau(\tilde{x})\, \omega'(\tilde{y}) \rangle,$$

where $x, y \in M_m$, $b \in (\pi')^{-1}(m)$, and $\tilde{x}, \tilde{y}$ are lifts to b. $H_z(x, y)$ is clearly independent of the choices involved and is symmetric since

$\tau\omega' = 0$. Hence, there is a corresponding symmetric transformation S_z on M_m given by

$$H_z(x, y) = \langle S_z x, y \rangle.$$

Now let T be the linear transformation field associated with the difference form τ, that is, if $x \in M_m$, $y \in N_{I(m)}$, then

$$T_x y = b\tau(\bar{x})\, b^{-1}(y),$$

where $b \in (\pi')^{-1}(m) \subset F(N, M)$ and $\bar{x}$ is a lift of x to b. In particular, $T_x(M_m) \subset M_m{}^\perp$, $T_x(M_m{}^\perp) \subset M_m$. If $y \in M_m$, then $T_x y = b\tau(\bar{x})\omega'(\bar{y})$. Now

$$\langle S_z x, y \rangle = H_z(x, y)$$
$$= -\langle z, b\tau(\bar{x})\, \omega'(\bar{y}) \rangle$$
$$= -\langle z, T_x y \rangle$$
$$= \langle T_x z, y \rangle$$

since $\tau(\bar{x})$ is skew-symmetric. Thus, for $x \in M_m$, $z \in M_m{}^\perp$,

$$S_z x = T_x z.$$

We have the following interpretation of T_x in terms of parallel translation. In fact, the connexions ϕ and ψ in $F_M(N)$ give rise to distinct parallel translations of the fibre, that is, of $N_{I(m)}$, $m \in M$. T is an infinitesimal measure of the difference.

Let $x \in M_m$, X a vector field on M defined in a neighborhood of m and such that $X(m) = x$. Let Y be a vector field of tangents to N defined on M, that is, $Y(m') \in N_{I(m')}$. Finally, let D and E denote covariant differentiation with respect to the connexions ϕ and ψ, respectively. Then we have

Proposition 1. $T_X Y(m) = (D_X Y - E_X Y)(m)$.

Proof. As in 6.4.1(iii), define a function $f_Y : F_M(N) \to R^f$ by: $f_Y(b) = b^{-1} Y(\pi(b))$. Let $\check{X}$ and $\tilde{X}$ be the ϕ and ψ horizontal lifts, respectively, of X to $F_M(N)$. Then from 6.4.1(iii),

$$(E_X Y)(m) = b((\tilde{X} f_Y)(b)),$$
$$(D_X Y)(m) = b((\check{X} f_Y)(b)),$$

where $\pi(b) = m$. But

$$
\begin{aligned}
(\tilde{X} - \bar{X}) f_Y(b) &= (\tilde{X} - \bar{X})(b) f_Y \\
&= \lambda \phi (\tilde{X} - \bar{X})(b) f_Y \\
&= \lambda \tau(\tilde{X})(b) f_Y \\
&= -\tau(\tilde{X}(b)) f_Y(b) \qquad \text{[lemma 5.5(ii)]} \\
&= -\tau(\tilde{X}) \, b^{-1} Y(m),
\end{aligned}
$$

so

$$
\begin{aligned}
T_X Y(m) &= b\tau(\tilde{X}(b)) \, b^{-1} \, Y(m) \\
&= b((\tilde{X} - \bar{X}) f_Y)(b) \\
&= (D_X Y)(m) - (E_X Y)(m). \qquad \text{QED}
\end{aligned}
$$

We give another interpretation of the transformation $T_x y$. Let $x \in M_m$, $y \in M_m^\perp$. Let γ be a curve with $\gamma(0) = m$ and $\gamma_*(0) = x$. For fixed t, let $Z(t)$ be the ψ parallel translate of y along γ to $\gamma(t)$, $Z(t) \in M_{\gamma(t)}^\perp$, and then let $Y(t)$ be the ϕ parallel translate of $Z(t)$ back along $I \circ \gamma$ to $I(m)$, so $t \to Y(t)$ is a curve in $N_{I(m)}$.

Proposition 2. $Y'(0) = T_x y$ $[^{(\prime)}$ denotes differentiation in the vector space $N_{I(m)}$.$]$

Proof.   Since Z is ψ parallel along γ, we have $E_x Z = 0$. But $D_x Z = Y'(0)$, so the result follows from proposition 1.

Problem 6.   (a) If Y_0 is a vector field along curve γ_0 in M and $Y = dI(Y_0)$ is the corresponding vector field along $\gamma = I \circ \gamma_0$ in N, then Y_0 is parallel (with respect to ϕ_0) if and only if $D_{\gamma_*} Y$ is in the normal bundle $\perp(M)$.

(b) Hence, γ_0 is a geodesic in M if and only if $D_{\gamma_*} \gamma_*$ is in $\perp(M)$.

10.5 Curvature and the Second Fundamental Form

We now apply this interpretation of the second fundamental form to our previous expression for curvature $K_0(x, y)$ to draw some immediate conclusions.

Recall that for orthonormal $x, y \in M_m$,

$$
K_0(x, y) = K(x, y) + \langle b[\tau(\bar{x}), \tau(\bar{y})] \, \omega'(\bar{x}), y \rangle,
$$

where $\pi(b) = m$ and $\bar{x}$, $\bar{y}$ are $(\phi_d + \phi_e)$ horizontal lifts of x, y to $b \in F(N, M)$.

Now

$$[\tau(\bar{x}), \tau(\bar{y})] \, \omega'(\bar{x}) = \tau(\bar{x}) \, \tau(\bar{y}) \, \omega'(\bar{x}) - \tau(\bar{y}) \, \tau(\bar{x}) \, \omega'(\bar{x})$$
$$= b^{-1}(T_x T_y x - T_y T_x x).$$

Note that similar arguments give

$$R_{0xy} = (PR_{xy} + T_x T_y - T_y T_x) \, |_{M_m} \, ,$$

where P is the orthogonal projection $N_{I(m)} \rightarrow M_m$.

Lemma 1.   Let $z_1 , ..., z_e$ be an orthonormal basis of $M_m^{\perp}$ and let $H_i = H_{z_i}$. Let x, y be an orthonormal pair in M_m . Then

$$K_0(x, y) = K(x, y) + \sum_i \left(H_i(x, x) \, H_i(y, y) - H_i(x, y)^2\right).$$

Thus, the induced curvature is the curvature of N plus the sum of the "squares of areas" with respect to the second fundamental form.

Proof.

$$\langle T_y x, z_i \rangle = -H_i(x, y),$$

so

$$T_y x = - \sum_i H_i(x, y) \, z_i.$$

Hence,

$$T_x T_y x = - \sum_i H_i(x, y) \, T_x z_i \, .$$

Now

$$\langle T_x z_i \, , y \rangle = H_i(x, y).$$

Therefore,

$$\langle T_x T_y x, y \rangle = - \sum_i H_i(x, y)^2.$$

Similarly,

$$T_x x = - \sum_i H_i(x, x) \, z_i \, ,$$

$$\langle T_y z_i \, , y \rangle = H_i(y, y),$$

so

$$- \langle T_y T_x x, y \rangle = \sum_i H_i(x, x) \, H_i(y, y).$$

Theorem 1.   Let M be immersed in N by I, let σ be a curve of M such that $I \circ \sigma$ is a geodesic of N. Let P be any plane section of M tangent to σ. Then $K_0(P) \leqslant K(P)$.

If we let Y be the ϕ_0 parallel vector field along σ such that P is spanned by $x = \sigma_*(0)$ and $y = Y(0)$, then equality holds between $K_0(P)$ and $K(P)$ if and only if $D_x dI(Y) = 0$. In particular, $K_0(Y, \sigma_*) = K(Y, \sigma_*)$ if and only if $dI(Y)$ is parallel in N.

Proof.   We may assume that Y and σ_* have been normalized, $\langle Y, Y \rangle = \langle \sigma_*, \sigma_* \rangle = 1$, $\langle Y, \sigma_* \rangle = 0$. The local arc length minimizing properties of geodesics (or problem 6) imply that σ is a geodesic in M. Hence, σ_* and $dI(\sigma_*)$ are parallel in M and N, respectively; so $T_{\sigma_*}\sigma_* = 0$ by proposition 1. Let z_i, H_i be as above, so $H_i(\sigma_*, \sigma_*) = 0$ for every i, and

$$K_0(Y, \sigma_*) - K(Y, \sigma_*) = - \sum_i H_i(Y, \sigma_*)^2 \leqslant 0.$$

This proves the first assertion.

It also shows that $K_0(P) = K(P)$ if and only if $H_i(y, x) = 0$ for every i. But

$$H_i(y, x) = - \langle z_i, T_x y \rangle,$$

so this is equivalent to $T_x y = 0$, or, by proposition 1,

$$D_x Y = E_x Y = 0,$$

since Y is parallel in M along σ.

Remark.   The case of the above theorem for dim $M = 2$ is known as *Synge's theorem* [73].

Let M be a submanifold of a Riemannian manifold N with the induced Riemannian structure. Then M is called a *totally geodesic submanifold* if every geodesic of M is a geodesic of N.

Theorem 2. M is a totally geodesic submanifold of N if and only if its second fundamental form vanishes identically.

Proof.   The vanishing of the second fundamental form H is clearly equivalent to the vanishing of the difference form τ, which means that the bundle $F(N, M)$ is situated "horizontally" in the bundle $F_M(N)$. If this is true, then parallel translation with respect to N and M coincide, and hence every parallel vector field in M is parallel in N, which proves that geodesics in M are also geodesics in N. Hence, M is totally geodesic in N.

Conversely, if M is totally geodesic, then every $x \in M_m$ is tangent to a curve σ which is a geodesic in both M and N. Thus by the same argument as in the proof of theorem 1, $H_i(x, x) = 0$ for all i, x. Hence, since H_i is symmetric, $H_i = 0$, all i, that is, $H = 0$. QED

Problem 7. Prove the following formula for the curvature transformations of an immersed manifold. Let z_i, H_i be as above, S_i the symmetric linear transformation such that $H_i(x, y) = \langle S_i x, y \rangle$, all $x, y \in M_m$. Then

(a) $\qquad R_{0xy}w = PR_{xy}w + \sum_i (\langle S_i x, w \rangle S_i y - \langle S_i y, w \rangle S_i x).$

If we view R_0, R as maps of bivectors into bivectors, $G^2{}_m \to G^2{}_m$, this becomes

(b) $\qquad R_0(xy) = P_2 R(xy) + \sum_i (S_i x)(S_i y),$

or

$$R_0 = P_2 R + \sum_i S_{i2}.$$

Here, P is the projection $N_{I(m)} \to M_m$, and if A is a linear transformation, A_2 is the extension to bivectors given by $A_2(xy) = (Ax)(Ay)$.

Problem 8. A submanifold M of N is *geodesic at m* if every geodesic of M through m is also a geodesic of N.

(a) If N has constant curvature show that every submanifold which is geodesic at a point is totally geodesic.

(b) Conversely, if every submanifold which is geodesic at a point is totally geodesic, then N satisfies the hypothesis of Schur's theorem, problem 9.3, so has constant curvature (dim $N > 2$) [17, pp. 232-233].

10.6 The Local Gauss Map

Let U be a Riemannian normal coordinate neighborhood in M at a point $m \in M$ (see 6.3.2). Let $\perp(U)$ be the restriction of the normal bundle of M to U. Then a map $G_U \colon \perp(U) \to M_m + M_m{}^\perp$ is defined by: if $z \in \perp(U)$, then $G_U(z) = $ parallel translate in N of z back along the geodesic ray in U from m to the base point of z.

G_U is called the *Gauss map with respect to U.*

Remark. If $N = R^j$, then parallel translation in N is independent of path, so that G may be defined globally on $\perp(M)$, and it will agree with G_U on $\perp(U)$. More generally, this works whenever N has a trivial holonomy group.

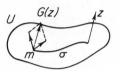

Fig. 35.

Notice that $\perp(M)$, as a bundle associated to $F(N, M)$, has a connexion corresponding to $\phi_d + \phi_e$ (5.4).

Proposition 3. Let $x \in M_m$, $z \in M_m{}^\perp$, and let $\tilde{x}$ be a horizontal lift of x to z in $\perp(U)$. Then $dG_U(\tilde{x})$ may be identified with an element y of $M_m + M_m{}^\perp$, and $y = T_x z$.

Proof. Let σ be the ray in U with tangent x at m. If $\tilde{\sigma}$ is the horizontal lift of σ to $F(N, M)$, then the curve

$$t \to \tilde{\sigma}(t)\,\tilde{\sigma}(0)^{-1}z$$

in $\perp(U)$ is tangent to $\tilde{x}$. Composing G_U with this curve gives us the curve

$$t \to \bar{\sigma}(0)\,\bar{\sigma}(t)^{-1}\tilde{\sigma}(t)\tilde{\sigma}(0)^{-1}z,$$

where $\bar{\sigma}$ is the horizontal lift of σ to $F_M(N)$. By construction, the tangent to this curve is $dG_U(\tilde{x}) = y$, but by proposition 2, $y = T_x z$.

QED

Corollary 1. If $x, y \in M_m$, $z \in M_m{}^\perp$, $\tilde{x}$ any lift of x to z in $\perp(U)$, then $H_z(x, y) = \langle dG_U(\tilde{x}), y\rangle$, where again we make an identification.

Proof. If $\tilde{x}$ is a horizontal lift, then this follows from the proposition above. If $\tilde{x}$ is not horizontal, then let x' be the horizontal lift, and notice that $dG_U(\tilde{x} - x')$ is tangent to $M_m{}^\perp$, so that its inner product with y is zero. This follows since $\tilde{x} - x'$ is tangent to $M_m{}^\perp$ and G_U is the identity on $M_m{}^\perp$. QED

10.7 Hessians of Normal Coordinates of N

Let M be any manifold, f a real-valued C^∞ function on M, $m \in M$. f has a *critical point* at n if $df_m = 0$. If m is a critical point of f, then the *Hessian of f at m*, H_f, is a bilinear function on M_m defined as follows. If $x, y \in M_m$, X a C^∞ vector field such that $X(m) = x$, then

$$H_f(x, y) = y(Xf).$$

Since $df_m = 0$, it is easy to verify that $H_f(x, y)$ does not depend on the choice of X and further that H_f is symmetric.

Problem 9. Verify these facts about H_f without using coordinates.

We now return to an immersion $I : M \to N$. The second fundamental form will be related to the Hessians of certain functions on M (see theorem 3 below).

Let V be a normal coordinate neighborhood in N at $n = I(m)$, and let $v_1, ..., v_f$ be normal coordinate functions on V so that $V_i(n) = (\partial/\partial v_i)(n)$ are an orthonormal basis of N_n. Let $u_i = v_i \circ I$. Then a linear combination $u = \sum_i a_i u_i$ has a critical point at m if and only if

$$\sum_i a_i V_i(n) \in M_m{}^\perp,$$

that is, $du = \sum_i a_i \, du_i$ annihilates M_m.

For simplicity we shall assume that $u = u_1$ has a critical point at m. We shall calculate $H_u(x, x)$ for all $x \in M_m$, hence determining H_u.

Let σ be the geodesic with $\sigma_*(0) = x$, and we shall use σ_* in the definition of $H_u(x, x)$. Let X be an extension of σ_* to a neighborhood of m, and let Y be an extension of $dI\,X$ to a neighborhood of n. Let the coordinate expression for Y be $Y = \sum_i g_i V_i$. But

$$Xu = \left(\sum_i g_i V_i v_1\right) \circ I = g_1 \circ I,$$

so

$$H_u(x, x) = xXu = x(g_1 \circ I) = dI(x)\,g_1 = yg_1,$$

where $y = Y(n)$.

By proposition 1 we have

$$
\begin{aligned}
T_x x &= D_x X - E_x X \\
&= D_x X \qquad \text{since } \sigma \text{ is a geodesic in } M. \\
&= D_y Y \\
&= D_y \left(\sum_i g_i V_i \right) \\
&= \sum_i (y g_i)\, V_i(n) + D_y \sum_i g_i(n)\, V_i \\
&= \sum_i (y g_i)\, V_i(n).
\end{aligned}
$$

This last step, $D_y \sum_i g_i(n)\, V_i = 0$, is true because the v_i are normal coordinates in N so that $\sum_i g_i(n)\, V_i$ is the tangent field to the geodesic in the direction of y.

Thus

$$
\begin{aligned}
H_u(x, x) &= y g_1 \\
&= \langle T_x x,\, V_1(n) \rangle \\
&= -H_z(x, x), \qquad z = V_1(n).
\end{aligned}
$$

We have proved:

Theorem 3.   Let $I : M \to N$ be an immersion, $m \in M$, V a normal coordinate neighborhood of N around $n = I(m)$, $u = \sum a_i u_i$, where $u_i = v_i \circ I$ are the normal coordinates pulled back to M. Assume $z = \sum a_i V_i(n) \in M_m^{\perp}$. Then u has a critical point at m and its Hessian form is the negative of the second fundamental form H_z.

If $N = R^f$, then the normal coordinates may be taken on all of N, and u is a linear function plus a constant pulled back to M. The Gauss map, defined on all of $\perp(M)$, may be considered as a map of $\perp(M)$ into N, identifying N_n with N, and G is independent of n. For any $n \in R^f$, we get a corresponding linear function u_n, $u_n(n') = \langle n', n \rangle$, on R^f. If $G(z) = n$, $z \in M_m^{\perp}$, then the normal coordinate combination corresponding to z is just a constant plus u_n. Hence we have:

Corollary 2.   If $I : M \to R^f$ is an immersion, $z \in M_m^{\perp}$, $G(z) = n$, then the following three forms are equivalent:

(a) H_z, the second fundamental form of z,

(b) the negative of the Hessian of $u_n \circ I$,

(c) the form $(x, y) \rightarrow \langle dG(\bar{x}), dI\, y \rangle$,

where $\bar{x}$ is a lift of x to $\bot(M)_z$.

Problem 10. Suppose that M is a submanifold in the neighborhood of $n \in N$ which is given as the locus of equations $g_i = 0$, $i = 1, ..., e$, where the g_i are real-valued C^∞ functions on N such that $g_i(n) = 0$ and $dg_i(n)$ are linearly independent. Let v_i be normal coordinates at n as above, and let $M' = V \cap \exp_n(M_n)$, where $\exp_n$ is the exponential function of N. Let

$$ g = \sum_i a_i g_i , \qquad dg(n) = \sum_i b_i\, dv_i(n), \qquad u = \sum_i b_i u_i , \qquad f = g\,|_{M'} . $$

Show that $H_f = -H_u$, both being defined on $M_n = M_n{}'$.

Problem 11. If $S_1 , ..., S_e$ are symmetric $d \times d$ matrices, then show that $R^d \rightarrow R^f$ given by $x \rightarrow (x,\ -\langle S_1 x, x \rangle, ..., -\langle S_e x, x \rangle)$ is an imbedding such that $S_1 , ..., S_e$ are the matrices of the second fundamental form with respect to the basis $D_i(0)$, $i \leqslant d$, of $R^d{}_0$, and basis $D_i(0)$, $i > d$, of the normal space (only at this one point).

Problem 12. Let $f = u_1 u_2 + u_3 u_4$, $g = u_1 u_3 + u_2 u_4 : R^4 \rightarrow R$. Then the intersection of $f^{-1}(1) \cap g^{-1}(0)$ with a neighborhood of $(1, 1, 0, 0)$ is a 2-dimensional submanifold of R^4. Find the curvature of this submanifold at the point $(1, 1, 0, 0)$.

10.8 A Formulation of the Immersion Problem

In what follows we shall be concerned with giving sufficient conditions that a map of M into N be an isometric immersion, and also giving sufficient structures from which such maps can be obtained. These sufficient conditions are also some of the necessary conditions from Sections 10.2 and 10.3, which involve the bundle of adapted frames. Since we wish to give conditions which depend mainly on the intrinsic structures of M and N, we must somehow construct the bundle of adapted frames *a priori*. If M is diffeomorphic to R^d this is no problem because all bundles over M will then be trivial, and the bundle of adapted frames will be equivalent to the product bundle $M \times O(d) \times O(e)$. If M is not so trivial, then the bundle of adapted frames will not necessarily be unique as a bundle; however, if there is to be a solution to our problem we must certainly be able to immerse M differentiably into N. If we combine the normal frames of such an immersion with the frames of M from the given Riemannian metric (not the metric induced by the immersion) we get a bundle B with group $G = O(d) \times O(e)$ which can serve as a model for our bundle of adapted frames. We assume henceforth that we have this bundle B.

If we had the isometric immersion $I : M \to N$, then we would also have the corresponding immersion $I' : B \to F(N)$. I' is a bundle map, in that it takes fibres into fibres and commutes with the action of elements of G. If we are given I', we can recover I by projecting into M and N. The approach to the problem of finding a suitable I' will be stated in terms of its graph, which is a submanifold of $P = B \times F(N)$. We note that G acts on P by means of the diagonal imbedding of G in $G \times O(f)$.

Proposition 4. A submanifold Q of P is the graph of a bundle map $I' : B \to F(N)$ if and only if Q is invariant under the action of G and the projection of Q onto B is a diffeomorphism.

Proof. If p_1 and p_2 are the projections of P onto B and $F(N)$, then the formula for I' is $I' = p_2(p_1 \mid_Q)^{-1}$. The proposition is a simple verification.

The solder forms of $F(M)$ and $F(N)$ can be pulled back to forms on P, the former through the intermediate bundle B, the latter directly by p_2. We shall denote them by θ and ω, respectively; ω decomposes into its R^d valued and R^e valued parts, $\omega = \omega' + \omega''$.

Let $\pi_1 : B \to M$ and $\pi_2 : F(N) \to N$ be the bundle projections. The properties of θ, ω we shall need are that for $t \in P_p$

$$\| d\pi_1 \, dp_1(t) \| = \| \theta(t) \| \qquad \text{and} \qquad \| d\pi_2 \, dp_2(t) \| = \| \omega(t) \| .$$

Theorem 4. A submanifold $j : Q \to P$, j the inclusion map, which is the graph of a bundle map I', has an isometric immersion for the induced map $I : M \to N$ if and only if $j^*(\theta - \omega') = 0$ and $j^*\omega'' = 0$.

Proof. Since by hypothesis $p_1 \mid_Q$ is a diffeomorphism,

$$J = (p_1 \mid_Q)^{-1} : B \to P$$

is defined.

Consider the following commutative diagram:

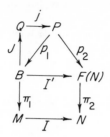

Suppose that $j^*(\theta - \omega') = 0$, $j^*\omega'' = 0$. Let $x \in M_m$, let $\bar{x}$ be a lift of x to $B : d\pi_1(\bar{x}) = x$. Then

$$\| dI(x) \| = \| d\pi_2 \, dI'(\bar{x}) \|$$
$$= \| d\pi_2 \, dp_2 \, dj \, dJ(\bar{x}) \|$$
$$= \| \omega' \, dj \, dJ(\bar{x}) \| + \| \omega'' \, dj \, dJ(\bar{x}) \|$$
$$= \| \theta \, dj \, dJ(\bar{x}) \| + 0$$
$$= \| d\pi_1 \, dp_1 \, dj \, dJ(\bar{x}) \|$$
$$= \| d\pi_1(\bar{x}) \|$$
$$= \| x \| .$$

That $j^*(\theta - \omega') = 0$ and $j^*\omega'' = 0$ follow from I being isometric has already been proved in Section 10.2.

We shall now give conditions under which such a submanifold Q can be constructed. Since these conditions will depend on having what is to be the second fundamental form and we wish to make our statement depend as much as possible on the intrinsic properties of M and N, it is desirable to make the hypothesized properties of the second fundamental form depend only on the curvature form of $F(M)$ and not also on that of $F(N)$. But the curvature form of $F(N)$ enters the formula for that of $F(M)$, so the only way is to make N have curvature which is independent of position and direction. Thus we *assume that N has constant curvature* k and hence that the curvature form of $F(N)$ pulls back by p_2 to $\Phi = k\omega\omega^t$ on P, where ω^t is the transpose of the column vector ω, so $\Phi_{ij} = k\omega_i\omega_j$ (see problem 9.27).

We adopt the convention that forms on B have a subscript 0, with the corresponding pullbacks to P having the subscript removed. Thus $\theta = p_1^*\theta_0$. The connexion form on $F(M)$ pulls back to B, by the projection $q : B \to F(M)$, to an $\mathfrak{o}(d)$-valued form ψ_0. Thus we let $\psi = p_1{}^*\psi_0$. Similarly for curvature, so the structural equation on P becomes $d\psi = -\psi^2 + \Psi$.

The connexion form on $F(N)$ pulls back by p_2 to a form ϕ on P. It decomposes into several blocks:

$$\phi' \qquad \mathfrak{o}(d)\text{-valued}$$

$$\tau \qquad d \times e \text{ matrix, } \mathfrak{t}_1\text{-valued}$$

$$-\tau^t \qquad e \times d \text{ matrix}$$

and

$$\phi'' \qquad \mathfrak{o}(e)\text{-valued}.$$

The structural equations on $F(N)$ then give

$$d\omega' = -\phi'\omega' + \tau\omega''$$

$$d\omega'' = -\tau^t\omega' + \phi''\omega''$$

$$d\phi' = -\phi'^2 + \tau\tau^t + k\omega'\omega'^t$$

$$d\tau = -\phi'\tau - \tau\phi'' + k\omega'\omega''^t$$

$$d\phi'' = -\tau^t\tau - \phi''^2 + k\omega''\omega''^t.$$

Let M be connected and simply connected, N complete.

Theorem 5. Let σ_0 and ψ_0'' be 1-forms on B such that $\psi_0 + \psi_0''$ is a connexion form with ψ_0'' the $\mathfrak{o}(e)$-valued part, σ_0 is horizontal, equivariant $\mathfrak{t}_1$-valued, and satisfying the conditions

(1) $$\Psi_0 = \sigma_0\sigma_0{}^t + k\theta_0\theta_0{}^t$$

(2) $$\sigma_0{}^t\theta_0 = 0$$

(3) $$d\sigma_0 = -\psi_0\sigma_0 - \sigma_0\psi_0''$$

(4) $$d\psi_0'' = -\psi_0''^2 - \sigma_0{}^t\sigma_0.$$

Then the codistribution on P spanned by $\theta - \omega'$, ω'', $\psi - \phi'$, $\sigma - \tau$, and $\psi'' - \phi''$ is involutive and every maximal connected integral manifold is a component of a manifold Q satisfying the hypothesis of theorem 4, and so gives an isometric immersion of M into N.

Proof. That this codistribution is involutive follows from the conditions (1)-(4) and the structural equations. For example,

$$
\begin{aligned}
d(\psi - \phi') &= -\psi^2 + \Psi + \phi'^2 - \tau\tau^t - k\omega'\omega'^t \\
&= -(\psi - \phi')\psi - \phi'(\psi - \phi') + \sigma\sigma^t + k\theta\theta^t - \tau\tau^t - k\omega'\omega'^t \\
&= -(\psi - \phi')\psi - \phi'(\psi - \phi') + (\sigma - \tau)\sigma^t + \tau(\sigma^t - \tau^t) \\
&\quad + k(\theta - \omega')\theta^t + k\omega'(\theta^t - \omega'^t).
\end{aligned}
$$

The other four exterior derivatives are found to be in the ideal generated by the five 1-forms in much the same way. (Actually it is the components of these vector-valued forms that should be considered, but the justification for using the whole forms in this way is trivial.)

We first show that the dimension of the codistribution is constant and complementary to the dimension of B. This follows from the fact that on the subspace of tangents to the factor $F(N)$ of P (that is, the kernel of dp_1) the five forms become $-\omega'$, ω'', $-\phi'$, $-\tau$, $-\phi''$, and these are a parallelization of that subspace, except for trivial duplications ($\phi_{ij}' = -\phi_{ji}'$). Thus the 5-tuple ($\theta - \omega'$, ω'', $\psi - \phi'$, $\sigma - \tau$, $\psi'' - \phi''$) is a linear map of each tangent space P_p onto $R^l + \mathfrak{o}(f)$ (more or less), so the kernel has constant dimension the same as B.

By this same argument the kernel of the linear map is complementary to the kernel of dp_1, so dp_1 is 1-1 and onto on the distribution in question. Hence the integral submanifolds will be mapped locally diffeomorphically into B.

Since $\psi - \phi'$ and $\psi'' - \phi''$ annihilate the tangents to the orbits of the diagonal action of G, a maximal integral submanifold Q_0 will contain a component of such an orbit whenever it contains a point. Moreover, the equivariance of the forms in the distribution guarantees that an element of G takes an integral manifold into an integral manifold. Thus $Q = Q_0 G$ will be an integral manifold invariant under G.

To show that $r = p_1 |_Q$ is a map onto B it suffices to prove that the horizontal lifts of geodesic segments in M can be lifted to Q: starting with a single point in B we can get all other points by moving along such curves and by the action of G. If γ_0 is the horizontal lift of a geodesic of M we will have

$$\theta_0(\gamma_{0*}) = x = \text{constant},$$

$$\psi_0(\gamma_{0*}) = 0, \qquad \psi_0''(\gamma_{0*}) = 0,$$

and if

$$\sigma_{0ij} = \sum_k s_{0ijk}\theta_{0k},$$

then

$$\sigma_{0ij}(\gamma_{0*}) = \sum_k x_k(s_{0ijk} \circ \gamma_0);$$

the functions $s_{0ijk} \circ \gamma_0$ are continuous and hence bounded, since the domain of γ_0 is a closed interval. The lift of γ_0 to P, which we shall call γ, must satisfy

$$\omega'(\gamma_*) = \theta(\gamma_*) = x, \qquad \omega''(\gamma_*) = 0,$$

$$\phi'(\gamma_*) = \psi(\gamma_*) = 0, \qquad \phi''(\gamma_*) = \psi''(\gamma_*) = 0,$$

$$\tau_{ij}(\gamma_*) = \sigma_{ij}(\gamma_*) = \sum_k x_k(s_{ijk} \circ \gamma).$$

Since it is only the component of γ which is in $F(N)$ whose existence is in question, we project γ_* onto $F(N)$, and we find that

$$dp_2\gamma_* = E(x) + \sum_{i,k \leqslant d, j > d} x_k(s_{ijk} \circ \gamma \circ p_2)F_{ij} = X.$$

Integral curves of such an X clearly exist by the condition of completeness of N, so the lift γ exists.

It remains to show that r is 1-1. To this end we note that Q is a principal bundle with group G and some base manifold M' and that $r : Q \to B$ is a bundle map. Since curves can be lifted from B to Q they can also be lifted from M to M'. Thus the induced map $M' \to M$ is a covering map, is 1-1 because M is simply connected, M' connected. This implies that r is also 1-1. QED

Remarks. (a) To find local imbeddings of M into N it suffices to construct forms such as σ_0, ψ_0'' on a cross section of M into B, because they can then be extended by equivariance to the remaining part of the fibre; instead of a cross section we can also use M itself.

(b) Equations (3) and (4) are the structural equations for σ_0 and ψ_0''. Equation (4) says that the $o(e)$-valued part of the curvature on B is $-\sigma_0{}^t\sigma_0$. To interpret (3) as a structural equation we consider the adjoint action of G on its orthogonal complement in $o(f)$. Then $\begin{pmatrix} 0 & \sigma_0 \\ -\sigma_0{}^t & 0 \end{pmatrix}$ has values in that G-module, is horizontal, and satisfies an invariance of the type in lemma 5.3, part (ii). Thus (3) may be restated as

$$D\sigma_0 = 0, \tag{3'}$$

where D is as in 5.3 for the connexion

$$\psi_0 + \psi_0''.$$

(c) Equations (1) and (2) are algebraic at each point. Cartan [16] has shown that they possess many solutions provided $f \geqslant \frac{1}{2} d(d + 1)$, so that (1) and (2) are solvable locally. The solutions which Cartan obtained are of a sort which he calls regular, and for analytic Riemannian manifolds this regularity implies the existence of local solutions of (3) and (4) also; this is the statement of the Cartan-Kähler theorem [15, 43] in our context. Analyticity is definitely required because the Cartan-Kähler theorem leans on the Cauchy-Kowalevsky theorem, which is true only in the analytic case [34, p. 81; 43].

Problem 13. The algebraic problem (1) and (2) is equivalent to finding second fundamental form matrices S_i, $i = 1, ..., e$, such that the curvature $K(x, y)$ of orthonormal vectors $x, y \in M_m$ is given by

$$K(x, y) = \sum_{i=1}^{e} (\langle S_i x, x \rangle \langle S_i y, y \rangle - \langle S_i x, y \rangle^2) + k.$$

Problem 14. Assuming Cartan's solution of (1) and (2), which only depends on the curvature identities, show that if R is a tensor (not a tensor field) satisfying those identities then there is a Riemannian manifold having R as curvature tensor at one point. (*Hint*: let $k = 0$ consider the imbedding of problem 11.)

The following problem gives an outline of the solution of (1) and (2), but with f larger by d than Cartan's solution.

Problem 15. Let $U = $ 4-linear functions on R^d satisfying the curvature identities, $S = $ symmetric $d \times d$ matrices, $P = $ symmetric 4-linear functions on R^d, and $T = $ 4-linear functions on R^d satisfying the identities: for $t \in T$, $x, y, z, w \in R^d$, $t(x, y, z, w) = t(y, x, z, w)$ $= t(z, w, x, y)$. P is a subspace of T; T may be considered to be the space of symmetric bilinear forms on $V = $ the symmetric part of $R^d \otimes R^d$. Let $D = d(d - 1)/2, E = D + d = $ dimension V.

When we speak of $t \in T$ being positive definite we mean it as a symmetric bilinear form on V.

Let $F : S \to T$ be defined by

$$F(s)(x, y, z, w) = \langle sx, y \rangle \langle sz, w \rangle,$$

and $G : T \to U$ by

$$G(t)(x, y, z, w) = t(x, z, y, w) - t(x, w, y, z).$$

Verify the following:

(a) If $s_1, ..., s_k \in S$, then $R = \Sigma_i\, G(F(s_i))$ is the formula which expresses the curvature tensor R in terms of second fundamental forms s_i for an imbedding in R^{d+k}.

(b) $\dim T = E(E+1)/2$, $\dim P = d(d^3 + 2d^2 + 3d + 2)/24$, $\dim U = d^2(d^2 - 1)/12 = \dim T - \dim P$, and G is linear with kernel P, hence G is onto U.

(c) A symmetric bilinear form is nonnegative semidefinite and of rank one if and only if its matrix may be written XX^t, where X is a column matrix.

(d) If we take as basis of V elements of the form $\frac{1}{2}(e_i \otimes e_j + e_j \otimes e_i)$, $i \leqslant j$, and index these basis elements by pairs $I = (i, j)$, $i \leqslant j$, then the matrix of $F(s)$ is $(s_I s_J)$, where $s_I = s_{ij}$ is the i, j entry of the $d \times d$ matrix s.

(e) Every nonnegative semidefinite symmetric bilinear form t on V may be written as a sum of $k = $ rank t of those of rank 1, $k \leqslant E$. Thus the set of sums $\Sigma_i\, F(s_i)$ is exactly the nonnegative semidefinite part of T.

(f) If we order the double indices I so that the repeated ones come first, and let $I_r = $ the identity matrix of order r, $K = $ the $d \times d$ matrix with all entries 1, then the bilinear form on V with matrix $\begin{pmatrix} I_d + K & 0 \\ 0 & I_D \end{pmatrix}$ is positive-definite and in P.

(g) Every coset $t + P$ contains a positive-definite element.

Problem 16.   The following abstracts part of the proof of theorem 5.

Let $F : M \to N$ be a differentiable map, where N is parallelized by 1-forms ω_i, $i = 1, ..., f$. Let p_1, p_2 be the projections of the product $M \times N$. Then the codistribution spanned by $p_1{}^* F^* \omega_i - p_2{}^* \omega_i$, $i = 1, ..., f$ is integrable and the graph of F is an integral submanifold.

Conversely, if forms θ_i, $i = 1, ..., f$, are given on M such that $p_1{}^* \theta_i - p_2{}^* \omega_i$ span an integrable codistribution, then an integral submanifold is locally the graph of maps of neighborhoods of M into N.

10.9 Hypersurfaces

A *hypersurface* is an immersed manifold in a space of one dimension higher; in the notation above $e = 1$. When the immersion is into R^f the curvature may be expressed in terms of one second fundamental form, which implies that each R_{xy} is either 0 or of rank 2, so the curvature transformations of a hypersurface in R^f are very special.

The possibilities for the bundle of adapted frames is also limited in the hypersurface case. In fact, the unit normal bundle must be either a connected double covering of M or two copies of M; thus B is either a double covering of $F(M)$ or two copies of $F(M)$. When M is simply connected only the two copy-case is possible. This makes the specialization of theorem 5 much simpler, since $\psi'' = 0$ is determined.

Theorem 6. Let M be a simply connected Riemannian manifold with curvature form $\Psi_0 = \sigma_0 \sigma_0{}^t + k\theta_0 \theta_0{}^t$, where σ_0 is an R^d-valued equivariant 1-form on $F(M)$ such that $D\sigma_0 = 0$, $\sigma_0 \theta_0{}^t = 0$. If N is a $(d + 1)$-dimensional complete manifold with constant curvature k, then M may be immersed in N.

The immersion I is uniquely determined by the specification of the second fundamental form σ_0, dI_m for one $m \in M$, and a choice of unit normal at M. This yields the uniqueness theorem:

Theorem 7. Let N be of constant curvature k and such that the group of isometries of N is transitive on $F(N)$ (e.g., $N = S^{d+1}$ or $N = R^{d+1}$). If $I_i : M \to N$, $i = 1, 2$, are isometric immersions of M as a hypersurface such that the second fundamental forms coincide, then there is an isometry J of N such that $JI_1 = I_2$.

Theorem 7 holds without the restriction that M be simply connected because we are given the existence of the immersions and thus know that the integral manifolds give the graphs of their prolongations uniquely. J is produced by forcing it to satisfy the condition at some single point: $dJ\, dI_{1m} = dI_{2m}$ and $dJ(z_1) = z_2$, where z_i is the chosen unit normal at m for I_i.

Problem 17. Let $I_i : M \to N$ be isometric imbeddings as hypersurfaces, $i = 1, 2$, such that the second fundamental forms agree and for some $m \in M$ there is an isometry $J : N \to N$ such that $dJ\, dI_{1m} = dI_{2m}$, and if z is a normal at m, $z' = dJ(z)$, then $S_{2z'} = S_{1z} \neq 0$, where S_2, S_1 are the second fundamental forms of I_1, I_2.

Prove that $JI_1 = I_2$. [*Hint*: Use problem 16. Alternately, if γ is a curve in M starting at m, $n = I_1(m)$, show that the development of $I_1 \circ \gamma$ into N_n depends only on the second fundamental forms along γ and the development of γ into M_m.]

Problem 18. Let $g : N \to R$ be differentiable. Show that

$$M = g^{-1}(0) \cap \{n \mid dg(n) \neq 0\}$$

is a hypersurface in N.

Problem 19. Let $g : R^f \to R$ be differentiable, and consider R^f provided with the Euclidean metric, giving the M of problem 18 the induced metric. Show:

(a) At any $m \in M$ there is a unique unit normal z_1 such that $dg(z_1) = a > 0$. These normal vectors form a differentiable field in $\bot(M)$, so M is orientable.

(b) If M' is the linear hypersurface of R^f through $m \in M$ which is $\bot z_1$, show that the second fundamental form H_1 of M at m is given by $H_1(x, y) = (1/a)\, H_{g'}(x, y)$, where $g' = g \mid_{M'}$, $x, y \in M_m$.

(c) If $x = \sum a_i D_i(m)$, $y = \sum b_i D_i(m)$, and we let $X = \sum a_i D_i$, $Y = \sum b_i D_i$, then the curvature in M of the plane of x and y is $K_0(x, y) = (1/a^2)\,\{x(Xg)\,y(Yg) - (x(Yg))^2\}$.

Problem 20. Use the formula for K_0 given in problem 19(c) to show that the curvature of the sphere of radius r, $S^d = g^{-1}(0)$, where $g = \sum u_i^2 - r^2 : R^{d+1} \to R$, is constant and equal to $1/r^2$.

Problem 21. A *ruled surface* is a surface in R^3 such that through every point there is a line (a *ruling*) in R^3 which is contained in the surface. Show that the curvature of a ruled surface is nonpositive.

Problem 22. Let γ be a ruling in a ruled surface, and assume that γ is the base curve of a C^∞ rectangle having rulings as longitudinal curves, parametrized by arc length.

(a) Use the fact that the associated vector field V is a Jacobi field in R^3 to show that it may be written $V = uA + B$, where A, B are parallel along γ in R^3. Moreover, if the initial transverse curve is chosen at the right point then we may have A, B, and γ_* mutually perpendicular.

(b) Use the fact that V is also a Jacobi field in the surface to conclude $E^2V = -KV$, where $K =$ curvature along γ, E is covariant derivative in the surface with respect to γ_*.

(c) Use the fact that EV, E^2V are the projections of DV, $D(EV)$ onto the tangent plane to the surface (proposition 1) to find E^2V explicitly in terms of $\langle A, A \rangle$ and $\langle B, B \rangle$ and thus find the formula for K along γ.

Problem 23. A surface in R^3 is *doubly ruled* if it contains two lines in independent directions through every point. Prove that a doubly ruled flat surface ($K = 0$) must be a plane.

Problem 24. Let M be immersed in R^d, $I : M \to R^d$, in such a way that the last coordinate $v_d = u_d \circ I$ is always positive on M. Let I' denote the first $d - 1$ coordinates of I, so $I = (I', v_d)$. Considering R^d as the subspace of R^{d+e} having the last e coordinates 0, we get an immersion J of $M \times S^e$, by defining

$$J(m, x) = (I'(m), v_d(m)\, x).$$

We say that $M \times S^e$ is immersed as a *manifold of partial rotation*. If M is a curve in R^2 we say $J(M \times S^e)$ is a *hypersurface of rotation*.

We give M the metric induced by I, $M \times S^e$ the metric induced by J.

(a) Show that $M \times \{x\} \subset M \times S^e$ is a totally geodesic submanifold for every $x \in S^e$ and is isometric to M.

(b) Show that $\{m\} \times S^e \subset M \times S^e$ is a totally geodesic submanifold if and only if m is a critical point of v_d, and in that case $\{m\} \times S^e$ is isometric to an e-sphere of radius $v_d(m)$.

(c) J is an imbedding if and only if I is also.

Problem 25. *Products of spheres as hypersurfaces.* If we let $M = S^d$ in problem 24, M imbedded in R^{d+1} as a translate of the usual S^d so as to have $v_d > 0$, then J gives us an imbedding of $S^d \times S^e$ as a hypersurface.

(a) Find the curvature $K(X, Y)$ in the cases:

(i) X, Y both tangent to S^d;

(ii) X tangent to S^d, Y tangent to S^e,

(iii) X, Y both tangent to S^e.

(b) Generalize this to an imbedding of a product of any number of spheres as a hypersurface, and find what can be said about totally geodesic submanifolds and curvature.

A special case is the usual imbedding of a torus as the surface of a bagel.

Problem 26.   Let M be an orientable hypersurface in R^{d+1}, N a submanifold contained in a bounded region of R^{e+1}. By considering $R^{d+1} \subset R^{d+e+1}$ we get a normal bundle to M which has fibres of dimension $e + 1$, and because M is orientable this normal bundle has $e + 1$ independent global cross sections.

(a) Show that there is a C^∞ positive function r on M such that the exponential map of R^{d+e+1} is a diffeomorphism on

$$\perp_r(M) = \{(m, x) \in \perp(M) \mid \| x \| < r(m)\}.$$

(b) Show that $M \times N$ may be imbedded in R^{d+e+1}.

(c) If M is compact show that r may be taken constant and hence the imbedding of $M \times N$ constructed in such a way as to have induced metric with formula

$$\langle s + t, s' + t' \rangle = \langle s, s' \rangle_n + c \langle t, t' \rangle_2,$$

where s, $s' \in M_m$, t, $t' \in N_n$, $\langle \, , \, \rangle_n$ is a metric on M which is a differentiable function of $n \in N$, $\langle \, , \, \rangle_2$ is the metric on N induced from R^{e+1}, and c is constant.

(d) If M, N are only immersed, then the procedure given will yield an immersion of $M \times N$. However, find an example where M is not imbedded, N is imbedded, and by taking r nonconstant we get an imbedding of $M \times N$.

Remark   When M is S^d the procedure in part (c) gives $N \times S^d$ as a manifold of partial rotation.

Problem 27.   If M is a hypersurface in R^{d+1} show that the curvature transformation, when viewed as a symmetric linear transformation of bivectors, is diagonal with respect to a basis of decomposable bivectors and has characteristic values $\lambda_i \lambda_j$, $i < j$, where λ_i are the characteristic values of the second fundamental form. The λ_i are classically the *principal curvatures*.

Problem 28. If M is a hypersurface of N, N having constant curvature, show the following relations between the second fundamental form and curvature at a point m:

(a) The curvature at m is constant and the same as that of N if and only if the second fundamental form has rank 0 or 1.

(b) The difference between the curvature transformations of M and N (restricted to M_m) has its range in a single two-dimensional subspace of M_m if and only if the second fundamental form has rank 2.

(c) If neither (a) nor (b), then the curvature determines the second fundamental form.

Problem 29. *Classical rigidity theorem.* Let N be homogeneous of constant curvature k, and let $I_i : M \to N$, $i = 1, 2$ be immersions of M as a hypersurface. The *type number* $t(m)$ of $m \in M$ is the rank of the second fundamental form at m; by problem 28, if $t(m) \geqslant 2$ then $t(m)$ is determined by the curvature independently of the immersion, but is only defined with respect to an immersion at points where M has constant curvature k.

Prove that if $t(m) > 2$ for every m, then there is an isometry J of N such that $JI_1 = I_2$.

Problem 30. If M is a hypersurface in R^{d+1}, then we define the *Gaussian curvature* of M to be the real-valued function k on M given by $k(m) = \det S_z$, where S_z is the second fundamental form transformation for a normal z at m.

(a) Show that k is well defined when d is even but only up to sign when d is odd.

(b) Show that $k(m)$ depends only on the metric on M and find a formula for it in terms of curvature. In particular, when $d = 2$, k coincides with the curvature.

Problem 31. Show that the curvature transformations at a point of a 3-dimensional Riemannian manifold may be realized as the curvature transformations at a point of a hypersurface in R^4.

Problem 32. Let M be a complete hypersurface of R^{d+1} such that the holonomy group of M is reducible, that is, there is a distribution V on M of dimension e, $e \neq 0$, d, which is invariant under parallel translation on any curve. The orthogonal complement $V^{\perp}$ is then another such distribution.

Assume that the set of points at which there is some nonzero curvature is dense in M. Show that:

(a) the second fundamental form is 0 on V or $V^{\perp}$, let us say V.

(b) the parallel translate of elements of V in M coincides with their parallel translate in R^{d+1}.

(c) the linear submanifold in R^{d+1} of dimension e with tangent space V_m, at any $m \in M$, is contained in M and the straight lines in this linear submanifold are geodesics in M.

(d) if R^{d-e+1} is the linear subspace of R^{d+1} perpendicular to V, then $M \cap R^{d-e+1} = N$ is a hypersurface of R^{d-e+1} and M is imbedded in R^{d+1} as a Riemannian product $N \times R^e \subset R^{d-e+1} \times R^e$.

CHAPTER 11

Second Variation of Arc Length

The first and second variations of arc length are considered and Synge's formula for the unintegrated second variation and its specializations are proved. The index form for general end points is defined, and after a treatment of the elementary properties of focal and conjugate points, the Morse index theorem for one fixed endpoint is proved. Minimum points and closed geodesics are discussed as well as various formulations of convexity. The chapter closes with a version of Rauch's comparison theorem and a number of consequences, including relations between curvature and volume [38, 50, 83].

11.1 First and Second Variation of Arc Length

We now take up the study of the relation between the lengths of the longitudinal curves of a rectangle and the associated vector field. If the base curve is a geodesic and the initial and final angles of the associated vector field satisfy certain reasonable conditions, then the derivative of the length with respect to the transverse parameter is zero. Hence, in this case it is the second derivative which determines how nearby curves compare in length with the base curve.

In general, these first and second derivatives of the lengths of longitudinal curves are given by differentiating the length integral under the integral sign with respect to the transverse parameter. Thus we get what are called the *first and second variations of arc length* along the curve of a rectangle in the form of integrals.

A *broken C^∞ rectangle* Q is a map $[a, b] \times [c, d] \to M$ such that there is a finite partition $u_0 = a < u_1 < \cdots < u_{n-1} < u_n = b$ of $[a, b]$ for which the restriction of Q to $[u_{i-1}, u_i] \times [c, d]$ is a C^∞ rectangle, $i = 1, ..., n$.

Problem 1. (a) Show that a broken C^∞ rectangle is continuous.

(b) If Q is a broken C^∞ rectangle show that $dQ(D_2(u, v))$ is well defined for every (u, v) in the domain of Q and defines a broken C^∞ rectangle in $T(M)$.

(c) On the other hand, $dQ(D_1(u_i, v))$ need not be well defined.

(d) If M has a Riemannian structure, then a broken C^∞ rectangle has a canonical lifting to $F(M)$ as defined in Section 8.1.

(e) If τ is a broken C^∞ curve in M and V is a broken C^∞ lift of τ to $T(M)$, then there exists a broken C^∞ rectangle Q having τ as base curve and $V(u) = dQ(D_2(u, 0))$ for $u \in [a, b]$, that is, V is the *vector field associated to* Q.

(f) If Q is a broken C^∞ rectangle, ω a form on M, then $Q^*\omega$ is not defined on the vertical lines $u = u_i$, although $Q^*\omega(D_2)$ may be defined continuously while $Q^*\omega(D_1)$ will have right and left limits, $Q^*\omega(D_1)(u_i^+, v)$ and $Q^*\omega(D_1)(u_i^-, v)$, respectively.

Now let Q be a broken C^∞ rectangle in M defined on $[a, b] \times [0, 1]$ with associated vector field V and longitudinal curves τ_y, $y \in [0, 1]$. Then the length of τ_y is

$$l(y) = \int_a^b \| dQ(D_1) \| (u, y) \, du.$$

So

$$l'(0) = \int_a^b D_2 \| dQ(D_1) \| (u, 0) \, du$$

and

$$l''(0) = \int_a^b D_2^2 \| dQ(D_1) \| (u, 0) \, du$$

are the first and second variations. The integrands $D_2 \| dQ(D_1) \| (u, 0)$ and $D_2^2 \| dQ(D_1) \| (u, 0)$ are called the *unintegrated first and second variations of arc length*.

Let $\tilde{Q}$ be a canonical lifting of Q to $F(M)$. Using the notation of Section 8.1, $\omega^0(D_1)$, $\omega^0(D_2)$, $\phi^0(D_1)$, etc., are functions with domains in R^2 and values in R^d or $\mathfrak{o}(d)$. $\omega^0(D_2)$ and $\phi^0(D_2)$ are defined and broken C^∞ on all of $[a, b] \times [0, 1]$, while $\omega^0(D_1)$ may be discontinuous on vertical lines $u = u_i$, but in any case has left and right limits and derivatives at these points. $\phi^0(D_1) = 0$ because the longitudinal curves of $\tilde{Q}$ are horizontal.

The structural equations applied to D_1, D_2 give

(1) $D_1 \omega^Q(D_2) - D_2 \omega^Q(D_1) = -\phi^Q(D_1) \, \omega^Q(D_2) + \phi^Q(D_2) \, \omega^Q(D_1)$

$\qquad\qquad = \phi^Q(D_2) \, \omega^Q(D_1);$

(2) $D_1 \phi^Q(D_2) - D_2 \phi^Q(D_1) = D_1 \phi^Q(D_2)$

$\qquad\qquad = -[\phi^Q(D_1), \phi^Q(D_2)] + \Phi^Q(D_1, D_2)$

$\qquad\qquad = \Phi^Q(D_1, D_2).$

The inner products of the longitudinal and transverse fields of Q are given by the corresponding inner products of $\omega^Q(D_1)$ and $\omega^Q(D_2)$.

Lemma 1.   Let Q be a C^∞ rectangle, and assume that the tangent to the base curve has constant length C. Then

$$l'(0) = -\frac{1}{C} \int_a^b \langle \omega^Q(D_2), D_1 \omega^Q(D_1) \rangle (u, 0) \, du$$

$$+ \frac{1}{C} \{ \langle \omega^Q(D_2), \omega^Q(D_1) \rangle (b, 0) - \langle \omega^Q(D_2), \omega^Q(D_1) \rangle (a, 0) \}.$$

Proof.   According to the last remark above,

$$D_2 \| dQ(D_1) \| (u, 0) = D_2 \langle \omega^Q(D_1), \omega^Q(D_1) \rangle^{1/2} (u, 0)$$

$$= \frac{1}{C} \langle D_2 \omega^Q(D_1), \omega^Q(D_1) \rangle (u, 0)$$

$$= \frac{1}{C} \langle D_1 \omega^Q(D_2) - \phi^Q(D_2) \, \omega^Q(D_1), \omega^Q(D_1) \rangle (u, 0)$$

$$= \frac{1}{C} \langle D_1 \omega^Q(D_2), \omega^Q(D_1) \rangle (u, 0)$$

$$= \frac{1}{C} D_1 \langle \omega^Q(D_2), \omega^Q(D_1) \rangle (u, 0)$$

$$- \frac{1}{C} \langle \omega^Q(D_2), D_1 \omega^Q(D_1) \rangle (u, 0);$$

we have used (1) and the fact that $\phi^Q(D_2)$ is skew-symmetric. Therefore,

$$l'(0) = -\frac{1}{C} \int_a^b \langle \omega^Q(D_2), D_1 \omega^Q(D_1) \rangle (u, 0) \, du$$

$$+ \frac{1}{C} \langle \omega^Q(D_2), \omega^Q(D_1) \rangle (u, 0)]_a^b,$$

which is the desired formula.

Corollary 1. If Q is a broken C^∞ rectangle having the tangent to the base curve a constant C in length, then

$$l'(0) = -\frac{1}{C}\int_a^b \langle \omega^Q(D_2), D_1\omega^Q(D_1)\rangle(u, 0)\, du$$

$$+\frac{1}{C}\sum_{i=1}^{n}\{\langle \omega^Q(D_2)(u_i, 0), \omega^Q(D_1)(u_i^-, 0)\rangle$$

$$-\langle \omega^Q(D_2)(u_{i-1}, 0), \omega^Q(D_1)(u_{i-1}^+, 0)\rangle\}.$$

Corollary 2. If the base curve of broken C^∞ rectangle Q is a geodesic (unbroken) and $\langle \omega^Q(D_2), \omega^Q(D_1)\rangle(b, 0) = \langle \omega^Q(D_2), \omega^Q(D_1)\rangle(a, 0)$, then $l'(0) = 0$. In particular, $l'(0) = 0$ if the transverse curves are all perpendicular to the base geodesic.

Proof. The base curve is a geodesic if and only if $\omega^Q(D_1)$ is C^∞ and $D_1\omega^Q(D_1)(u, 0) = 0$ for every u. Under the hypothesis given the integrand in corollary 1 is 0 and the sum telescopes to give

$$\langle \omega^Q(D_2), \omega^Q(D_1)\rangle(b, 0) - \langle \omega^Q(D_2), \omega^Q(D_1)\rangle(a, 0) = 0.$$

Corollary 3. Let N, P be submanifolds of M, and let us consider only broken C^∞ rectangles whose initial and final transversals are in

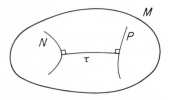

Fig. 36.

N and P. Then a curve τ has the property that for all such rectangles with base curve τ, $l'(0) = 0$ if and only if τ is a geodesic from N to P which is perpendicular to both N and P.

Proof. The idea is that we are able to get rectangles with sufficiently arbitrary $\omega^Q(D_2)$ that the formula for $l'(0) = 0$ will yield $D_1\omega^Q(D_1)(u, 0) = 0$, and then that τ is smooth at the breaks, perpendicular to N and P at the ends.

Let $\bar{\tau}$ be a horizontal lift to $F(M)$. If $r \neq u_i$, define a curve γ in R^d by $\gamma(u) = f(u) \, D\omega(\bar{\tau}_*)(u)$, where $f(u)$ is a non-negative C^∞ function which is positive at r and 0 at every u_i. Then $V = \bar{\tau}\gamma$ is a lift of τ to $T(M)$, so we may define $Q(u, v) = \exp_{\tau(u)} v V(u)$. It is easy to see that for this Q,

$$\langle \omega^Q(D_2), D_1 \omega^Q(D_1) \rangle (u, 0) = f(u) \, \| D_1 \omega^Q(D_1) \|^2 (u, 0),$$

and that the terms in the sum of corollary 1 are all 0. Thus $l'(0) = 0$ gives

$$\int_a^b f(u) \, \| D_1 \omega^Q(D_1) \|^2 (u, 0) \, du = 0,$$

which implies $D_1 \omega^Q(D_1)(r, 0) = 0$.

The terms in the sum are treated similarly. We show that $\langle t, \tau_*(u_i^+) - \tau_*(u_i^-) \rangle = 0$ for every $t \in M_{\tau(u_i)}$, $0 < i < n$, by parallel translating t along τ, multiplying the field so generated by a non-negative function positive near u_i but 0 outside a neighborhood of u_i, and defining Q as above.

The rectangles to show normality at the ends of τ may be taken to be those having longitudinal curves consisting of a short geodesic segment from $\gamma(v)$ to $\tau(a + \epsilon)$ and the segment $\tau|_{[a+\epsilon, b]}$, where γ is a curve in N with $\gamma(0) = \tau(a)$; the other end is treated likewise.

Problem 2. Show (independently of the arguments in Chapter 8) that a curve which minimizes distance between two points is a geodesic. (Classically, a geodesic was defined as a solution of the calculus of variations problem of finding minimal curves, with self-parallel condition $\nabla_{\gamma_*} \gamma_* = 0$ following as a consequence.)

Our applications of variational theory will only involve variation of a geodesic in a direction perpendicular to the geodesic at every point, that is, we shall only be using rectangles having a geodesic as base curve and the associated vector fields perpendicular to the base. As further justification for this assumption we prove the following remark which shows that for C^∞ rectangles satisfying the end conditions the assumption entails no loss at all. The remark is not true for *broken* C^∞ rectangles, however, unless the base geodesic is allowed broken linear reparametrizations.

Remark. If Q is a C^∞ rectangle having a geodesic base curve and with the initial and final transversals perpendicular to the base curve, then there is a partial reparametrization of Q for which the longitudinal curves are reparametrizations of those of Q and the associated vector field is perpendicular to the base.

Proof.   Let V be the associated vector field of Q, τ the base curve of Q, $f = \langle V, \tau_* \rangle$, so f is a real-valued function on $[a, b]$. By hypothesis $f(a) = f(b) = 0$. Define map $F : [a, b] \times [c, c + \epsilon] \to [a, b] \times [c, d]$ by

$$F(u, v) = \left(u - \frac{1}{k} vf(u), v \right), \qquad \text{where} \qquad \epsilon > 0$$

is such that the range of F is in $[a, b] \times [c, d]$, and $k = \langle \tau_*, \tau_* \rangle$. Then $Q \circ F$ is the desired rectangle.

Problem 3.   Complete the proof that the vector field associated to $Q \circ F$ is perpendicular to τ.

Henceforth we shall assume that all broken C^∞ rectangles have geodesic base and associated vector field perpendicular to the base. For convenience we also assume that the base curve is normalized so as to have unit tangent vectors, initial parameter value $a = 0$, and hence length $b = b - a$.

Lemma 2 (Synge's formula [86]).   The unintegrated second variation is given by

$$D_2{}^2 \| \omega^Q(D_1) \| = \| D_1\omega^Q(D_2) \|^2 - K(dQ(D_1), dQ(D_2)) \| \omega^Q(D_2) \|^2$$
$$+ D_1\{\langle \omega^Q(D_1), D_2\omega^Q(D_2)\rangle + \langle \omega^Q(D_1), \phi^Q(D_2)\, \omega^Q(D_2)\rangle\},$$

where all these functions are restricted to $[0, b] \times \{0\}$.

Proof.   The proof is by direct computation, always recalling that we are dropping the argument $(u, 0)$:

$$D_2{}^2 \| \omega^Q(D_1) \| = D_2(D_2\langle \omega^Q(D_1), \omega^Q(D_1)\rangle^{1/2})$$
$$= -\langle D_2\omega^Q(D_1), \omega^Q(D_1) \rangle^2 + D_2 \langle D_2\omega^Q(D_1), \omega^Q(D_1)\rangle,$$

since the factors appearing in the denominators are powers of $\langle \omega^Q(D_1), \omega^Q(D_1)\rangle(u, 0) = 1$.

Now using the first structural equation (1) and the fact that $\phi^Q(D_2)$ is skew-symmetric gives

(a) $D_2{}^2 \| \omega^Q(D_1) \| = -\langle D_1\omega^Q(D_2), \omega^Q(D_1)\rangle^2$
$$+ D_2\langle D_1\omega^Q(D_2), \omega^Q(D_1)\rangle.$$

The base curve τ is a geodesic, so $D_1\omega^Q(D_1) = 0$, hence,

$$\langle D_1\omega^Q(D_2), \omega^Q(D_1)\rangle^2 = (D_1\langle\omega^Q(D_2), \omega^Q(D_1)\rangle)^2 = 0,$$

because the associated field is perpendicular to τ.

(b)
$$D_2\langle D_1\omega^Q(D_2), \omega^Q(D_1)\rangle = \langle D_2D_1\omega^Q(D_2), \omega^Q(D_1)\rangle$$
$$+ \langle D_1\omega^Q(D_2), D_2\omega^Q(D_1)\rangle$$

(c)
$$\langle D_2D_1\omega^Q(D_2), \omega^Q(D_1)\rangle = \langle D_1D_2\omega^Q(D_2), \omega^Q(D_1)\rangle$$
$$= D_1\langle D_2\omega^Q(D_2), \omega^Q(D_1)\rangle$$

(d)
$$\langle D_1\omega^Q(D_2), D_2\omega^Q(D_1)\rangle = \langle D_1\omega^Q(D_2), D_1\omega^Q(D_2)$$
$$-\phi^Q(D_2)\,\omega^Q(D_1)\rangle$$

(e)
$$\langle D_1\omega^Q(D_2), \phi^Q(D_2)\,\omega^Q(D_1)\rangle = D_1\langle\omega^Q(D_2), \phi^Q(D_2)\,\omega^Q(D_1)\rangle$$
$$-\langle\omega^Q(D_2), (D_1\phi^Q(D_2))\,\omega^Q(D_1)\rangle$$

(f)
$$\langle\omega^Q(D_2), (D_1\phi^Q(D_2))\,\omega^Q(D_1)\rangle = \langle\omega^Q(D_2), \Phi^Q(D_1, D_2)\,\omega^Q(D_1)\rangle$$
$$= -K(dQ(D_1), dQ(D_2))\,\|\,\omega^Q(D_2)\,\|^2.$$

The desired result now follows by substituting (f)-(b) into their predecessors, finally into (a).

Corollary 1. Let V be the associated vector field along base geodesic τ, V' the covariant derivative with respect to τ_*, and let $\tilde{V}$ be the transverse vector field of Q.
Then

$$l''(0) = \int_0^b (\|\,V'\,\|^2(u) - K(V)\,\|\,V\,\|^2(u))\,du + \langle\tau_*, \nabla_V\tilde{V}\rangle]_0^b.$$

$(K(V) = K(V,\tau_*),$ cf. 9.4.)
This follows from the observation $V'(u) = \bar{\tau}(D_1\omega^Q(D_2)(u, 0))$ and $\nabla_V\tilde{V} = \tilde{Q}(D_2 + \phi^Q(D_2))\,\omega^Q(D_2)$ (cf. theorem 6.11).

Corollary 2. Let N and P be submanifolds, Q a rectangle with base geodesic τ perpendicular to N and P and with initial and final transversals in N and P. Then

$$l''(0) = \int_0^b (\|\,V'\,\|^2(u) - K(V)\,\|\,V\,\|^2(u))\,du$$
$$+ \langle S_{\tau_*(0)}V(0), V(0)\rangle - \langle S_{\tau_*(b)}V(b), V(b)\rangle$$

where S is the second fundamental form of the appropriate manifold and V is the associated vector field as before.

Proof.   The integral is the same as in corollary 1. By proposition 10.1, if $T_{V(0)}$ is the difference transformation of $M_{\tau(0)}$ given by N as a submanifold of M, then for vector field W on N

$$T_{V(0)}W(\tau(0)) = D_{V(0)}W - E_{V(0)}W.$$

But $E_{V(0)}W$ is perpendicular to $\tau_*(0)$, so for $W = \bar{V}$,

$$
\begin{aligned}
\langle \tau_*(0), D_{V(0)}\bar{V} \rangle &= \langle \tau_*(0), T_{V(0)}V(0) \rangle \\
&= -\langle T_{V(0)}\tau_*(0), V(0) \rangle \\
&= -\langle S_{\tau_*(0)}V(0), V(0) \rangle.
\end{aligned}
$$

Similarly for the other term.

Note that when N and P are points, the above formula for the second variation reduces to

$$l''(0) = \int_0^b (\| V' \|^2(u) - K(V) \| V \|^2(u))\, du$$

Corollary 3.   If in corollary 2 V is a Jacobi field, then

$$l''(0) = \langle S_{\tau_*(0)}V(0) - V'(0), V(0) \rangle$$
$$-\langle S_{\tau_*(b)}V(b) - V'(b), V(b) \rangle.$$

This follows from the Jacobi equation and integration by parts.

11.2 The Index Form

Let N and P be submanifolds of M, and let τ be a geodesic segment perpendicular to N and P at its ends $\tau(0)$ and $\tau(b)$. Let $\mathscr{L}$ be the linear space of broken C^∞ vector fields along τ which are perpendicular to τ and have their initial and final vectors tangent to N and P. The *index form* at τ is a bilinear form on $\mathscr{L}$ defined by: if $V, W \in \mathscr{L}$,

$$I(V, W) = \int_0^b (\langle V', W' \rangle(u) - \langle R_{\tau_* V}\tau_*, W \rangle(u))\, du$$
$$+ \langle S_{\tau_*(0)}V(0), W(0) \rangle - \langle S_{\tau_*(b)}V(b), W(b) \rangle.$$

Roughly speaking, if we consider the length function defined on curves from N to P, then corollary 3 of lemma 1 says that the critical points of this function are geodesics such as τ. The index form may be viewed as a natural generalization of the Hessian of a function at a critical point, since it is the symmetric bilinear form associated with the quadratic form $l''(0)$ of V in corollary 2 to lemma 2. Morse theory is concerned with this generalization.

If we observe that, except for points where V is not differentiable ($u_0 = 0, ..., u_n = b$),

$$\langle V', W' \rangle = \langle V', W \rangle' - \langle V'', W \rangle,$$

then we get another formula for $I(V, W)$,

$$I(V, W) = \sum_{i=1}^{n-1} \langle V'(u_i^-) - V'(u_i^+), W(u_i) \rangle$$

$$- \int_0^b \langle V'' + R_{\tau_* V^{\tau_*}}, W \rangle(u) \, du$$

$$+ \langle S_{\tau_*(0)} V(0) - V'(0), W(0) \rangle - \langle S_{\tau_*(b)} V(b) - V'(b), W(b) \rangle.$$

From this formula and application of techniques similar to that used in the proof of corollary 3 to lemma 1 it is clear what the null space of I is, that is, what conditions V must satisfy in order that $I(V, W) = 0$ for all $W \in \mathscr{L}$.

We single out the properties satisfied at each end.

Definition.   A Jacobi field V is an *N-Jacobi field* if it satisfies

   (i) V is perpendicular to the geodesic τ;
   (ii) $V(0) \in N_{\tau(0)}$;
   (iii) $S_{\tau_*(0)} V(0) - V'(0)$ is perpendicular to $N_{\tau(0)}$.

If N is a single point, $N = \tau(0)$, then (i), (ii), and (iii) reduce to $V(0) = 0$, $V'(0)$ is perpendicular to $\tau_*(0)$.

Problem 4.   Show that the N-Jacobi fields form a linear space of dimension $d - 1$, where $d =$ dimension M.

Theorem 1.   The null space of I consists of the intersection of the spaces of N-Jacobi fields and P-Jacobi fields.

We have already seen that a Jacobi field is characterized by the fact that it is the field associated to a rectangle having geodesics for longitudinal curves. The next theorem gives a similar characterization of N-Jacobi fields.

Theorem 2. V is an N-Jacobi field if and only if V is associated to a rectangle Q such that all of the longitudinal curves of Q are geodesics starting perpendicularly from N and parametrized by arc length.

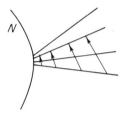

FIG. 37.

Proof.   By problem 4 it is sufficient to show that the Jacobi fields associated to such rectangles are N-Jacobi fields and have values at some $\epsilon > 0$ which equal the space normal to $\tau_*(\epsilon)$.

The rectangles in question can be factored through $\perp(N)$ by means of the exponential map $\perp(N) \to M$. In fact the longitudinal vector field $dQ(D_1) = X$ to such a rectangle Q gives a curve γ, $\gamma(v) = X(0, v)$, in $\perp_1(N)$, the unit normal bundle, and Q may be expressed as $Q(u, v) = \exp u\gamma(v)$.

Since $V(0) = dQ(D_2(0, 0))$ is the tangent to the projection of γ into N, (ii) is satisfied.

$$
\begin{aligned}
S_{\tau_*(0)} V(0) - V'(0) &= T_{V(0)} X(0) - D_{X(0)} V \\
&= D_{V(0)} X - E_{V(0)} X - D_{X(0)} V \\
&= \bar{Q}(0, 0)(D_2 \omega^Q(D_1) + \phi^Q(D_2)\, \omega^Q(D_1) \\
&\quad - D_1 \omega^Q(D_2) - \phi^Q(D_1)\, \omega^Q(D_2))(0, 0) - E_{V(0)} X \\
&= -E_{V(0)} X,
\end{aligned}
$$

by the first structural equation (1). But E is covariant derivative on N and the normal bundle, so the derivative of normal field $X(0, \cdot)$ is again a normal vector, which shows (iii). Moreover, to show (i) we need only show that V and V' are perpendicular to τ_* at one point, which we have already for $V(0)$, hence for $S_{\tau_*(0)}V(0)$. Thus it suffices to show $E_{V(0)}X$ is perpendicular to $\tau_*(0) = X(0)$. But

$$D_2 \langle X, X \rangle (0, 0) = 2 \langle E_{V(0)}X, X(0) \rangle = 0$$

since $\langle X, X \rangle = 1$.

We know that $\exp : \perp(N) \to M$ is nonsingular on the 0 cross section and thus also at $x = \epsilon\tau_*(0)$ for some $\epsilon > 0$. However, the tangents at x to rectangles in $\perp(N)$ of the form $(u, v) \to u\gamma(v)$, γ as above, fill up the tangent space $\perp(N)_x$. Consequently, the tangents at $\tau(\epsilon)$ of rectangles Q fill up the tangent space $M_{\tau(\epsilon)}$, and the normal parts, $V(\epsilon)$, fill up the space normal to $\tau_*(\epsilon)$.

Corollary. The range of $d \exp_x : \perp(N)_x \to M_p$, where $p = \exp x = \tau(u)$, is the space spanned by $\tau_*(u)$ and the values at u of N-Jacobi fields, $u \neq 0$.

Proof. Any tangent at x may be decomposed into a vector tangent to $\perp_u(N)$ and a vector tangent to the ray through x. The component tangent to $\perp_u(N)$ maps by $d \exp$ to the value of an N-Jacobi field, the component along the ray maps to a multiple of $\tau_*(u)$. ($\perp_u(N) = \{y \in \perp(N) \mid \|y\| = u\}$.)

We shall denote the quadratic form associated with I by I also, thus writing $I(V) = I(V, V)$. Since $I(V)$ is, for $V \in \mathscr{L}$, the second variation $l''(0)$ of a rectangle attached to V, we have immediately:

Theorem 3. If $V \in \mathscr{L}$ is such that $I(V) < 0$ then every neighborhood of τ contains shorter curves from a neighborhood of $\tau(0)$ in N to a neighborhood of $\tau(b)$ in P.

More generally, the dimension of a maximal subspace of $\mathscr{L}$ on which I is negative definite, the *index of I*, tells in how many independent directions τ can be pushed so as to shorten length, still obtaining a curve from N to P. It is shown below that the index is finite.

Problem 5. Let $M = R^d$ with the Euclidean metric and suppose that submanifolds N and P do not intersect. Let $K(N, P) = $ broken C^∞ curves from N to P, $H(N, P) = $ straight lines from N to P. Suppose that these curves are parametrized by *reduced arc length*, that is, they all have domain $[0, 1]$ and constant-length tangents. We make

$K(N, P)$ into a metric space by defining the distance between curves σ, τ to be

$$\rho(\sigma, \tau) = \max \rho(\sigma(u), \tau(u)) + \| \, |\sigma| - |\tau| \, \|.$$

Let $\phi : K(N, P) \to H(N, P)$ be the map which assigns to σ the line segment with the same end points as σ. Show that

(a) ϕ is homotopic to the identity via a homotopy which leaves $H(N, P)$ fixed. Thus $H(N, P)$ is a deformation retract of $K(N, P)$ and they have the same ordinary topological invariants.

(b) $H(N, P)$ is topologically the same as $N \times P$ and the length function L is differentiable when viewed as a function on $N \times P$.

(c) Vector fields V which are associated to rectangles with range in $H(N, P)$ are Jacobi fields in R^d and may be identified with tangents to $N \times P$.

(d) The critical points of L as a differentiable function on $N \times P$ are the line segments which are perpendicular to N and P at the ends.

(e) The Hessian of L is essentially the index form restricted to Jacobi fields in $\mathscr{L}$.

11.3 Focal Points and Conjugate Points [37, 57, 59, 60, 61, 73]

Let N be a submanifold of M, $\perp(N)$ the normal bundle. The exponential map of M by restriction gives a map $\exp : \perp(N) \to M$, which we have already seen to be a diffeomorphism in a neighborhood of the zero cross section. For $n \in N$, let $\perp(N)(n)$ be the fibre of $\perp(N)$

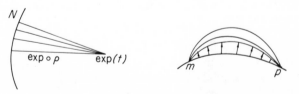

FIG. 38.

over n. Then $t \in \perp(N)(n)$ is a *focal point of N* if $d \exp$ is singular at t. If ρ is the ray from 0 to t in $\perp(N)(n)$, then $\exp(t)$ is called a *focal point of N along $\exp \circ \rho$*, which is, of course, a geodesic perpendicular to N. When N is a single point, say m, so $\perp(N) = M_m$, then a focal point is called a *conjugate point to m*. The *order* of a focal point is the dimension of the linear space annihilated by d exp.

Theorem 2 and its corollary show that it is equivalent to define focal points in terms of Jacobi fields as follows. If τ is a geodesic which starts perpendicular to N, then $\tau(b)$ is a focal point of N along τ if and only if there is an N-Jacobi field which vanishes at b. The order of $\tau(b)$ is the dimension of the space of such Jacobi fields. By theorem 1, the order is also the nullity of the index form at τ with end manifolds N and the single point $\tau(b)$.

The last statement seems to indicate how to generalize the concept of focal point to something which might be called a "focal submanifold." Presumably one would then aim to prove an "index theorem," as we shall below for the case of one end-manifold and a point, for the two end-manifold case. The purpose of such a theorem is to express the index of the index form in terms of orders of focal points (or focal submanifolds) in between the end manifolds. Such theorems have been formulated by both Morse [57] and Ambrose [1] for the two end-manifold case, but the difficulties in statement and proof are far greater than for the case we give, for which the formulation given by Morse is clearly the right one.

Problem 6. Show that conjugacy is a symmetric relation, that is, if m is conjugate to n along geodesic τ, then n is conjugate to m along $-\tau$, where $-\tau$ is the geodesic with initial tangent $-\tau_*(0)$.

Problem 7. Show that if τ is a geodesic segment perpendicular to N on which there are no focal points of N, then there is a neighborhood U of τ in M and a neighborhood V of $\tau(0)$ in N such that τ minimizes distance among curves in U which go from a point of V to $\tau(b)$ (cf. theorem 8.2).

Problem 8. Show that if M is complete and contains a point which has no conjugate points, then M is covered by R^d.

Problem 9. Show that if M has nonpositive curvature, then there are no conjugate points.

Problem 10. Determine the conjugate points and their orders for a point on a d-sphere of constant curvature.

Problem 11. Let $CP^d = S^{2d+1}/S^1$ have the metric induced by the metric on S^{2d+1} having curvature 1. Show that the conjugate points to a point in CP^d occur at distances $(2n + 1)\pi/2$ and $n\pi$, n an integer, with orders 1 and $2d - 1$, respectively. Use the fact that a Jacobi field attached to a family of horizontal geodesics in S^{2d+1} must project to a Jacobi field on CP^d.

Problem 12. In a way similar to that of problem 11, show that the conjugate points to a point in $QP^d = S^{4d+3}/S^3$, S^{4d+3} = unit sphere, occur at distances $(2n + 1)\,\pi/2$ and $n\pi$ with orders 3 and $4d - 1$, respectively.

11.4 The Infinitesimal Deformations

A useful technique for analyzing the index form is to break it into the sum of several index forms obtained by inserting intermediate manifolds of dimension $d - 1$ normal to τ. Since we will be dealing with several index forms, but all along segments of the same geodesic τ, we introduce the notation $I(T_1, T_2)$ to mean the index form with end manifolds T_1 and T_2. We shall also omit restriction symbols, that is, if $V \in \mathscr{L}$, the domain of $I(N, P)$, T_1 and T_2 are normal to τ at $\tau(u_1)$ and $\tau(u_2)$, and $I_1 = I(T_1, T_2)$, then we shall write $I_1(V)$ instead of $I_1(V\,|_{[u_1,u_2]})$.

If we choose the intermediate manifolds close enough together then any curve in some neighborhood of τ [by *neighborhood of* τ we shall mean curves which go from a neighborhood of $\tau(0)$ in N to a neighborhood of $\tau(b)$ in P, as well as lying in some open set containing τ] can be replaced by a unique shorter broken geodesic with breaks only at the intermediate manifolds. If T_i is placed at $\tau(u_i)$, $i = 1, ..., n$, $u_i < u_{i+1}$, then the requirement is that there be no focal point of N on $\tau\,|_{[0,u_i]}$, that there be no conjugate point of $\tau(u_i)$ on $\tau\,|_{[u_i,u_{i+1}]}$, $i = 1, ..., n - 1$, and that there be no focal point of P on $\tau\,|_{[u_n,b]}$. A curve σ sufficiently close to τ will intersect each T_i, and we let the first intersection be p_i; requiring σ to be even closer if necessary, there will be a unique shortest geodesic segment from p_1 to the neighborhood of $\tau(0)$ in N, a unique shortest geodesic segment from p_i to p_{i+1}, and a unique shortest geodesic segment from p_n to the neighborhood of $\tau(b)$ in P. Chaining these geodesic segments together gives a broken geodesic γ having breaks only on the intermediate manifolds T_i. The map $\phi : \sigma \to \gamma$ is a length-nonincreasing deformation of curves nearby τ into the much smaller collection of broken geodesics; in fact, the range of ϕ may be viewed as the product manifold

$T_1 \times T_2 \times \cdots \times T_n = T$, and the length function on curves as a differentiable function on T. τ is a critical point of the length function and the index form is the Hessian at τ.

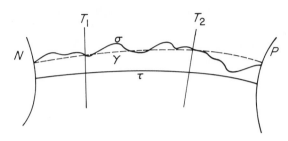

Fig. 39.

Detailed proofs of these facts are more proper to an exposition of Morse theory, so we content ourselves here with their infinitesimal versions.

Lemma 3.   Under the conditions prescribed on the choice of the u_i, for every choice of vectors $y_i \in \tau_*(u_i)^{\perp}$ there is a unique vector field $Y \in \mathscr{L}$ such that $Y(u_i) = y_i$, $Y|_{[u_i, u_{i+1}]}$ is a Jacobi field, and $Y|_{[0, u_1]}$, $Y|_{[u_n, b]}$ are N- and P-Jacobi fields, respectively.

The map $G : \Sigma_{i=1}^{n} \tau_*(u_i)^{\perp} \to \mathscr{L}$, $G(y_1, \ldots, y_n) = Y$, is a linear isomorphism.

Proof.   The linear transformation which assigns to an N-Jacobi field V its value $V(u_1)$ is 1-1 onto $\tau_*(u_1)^{\perp}$ because $\tau(u_1)$ is not a focal point of N. Thus $Y|_{[0, u_1]}$ exists and is unique; similarly for $Y|_{[u_n, b]}$.

For the same reason there is a unique $\tau(u_i)$-Jacobi field V_i and a unique $\tau(u_{i+1})$-Jacobi field W_i such that $V_i(u_{i+1}) = y_{i+1}$, $W_i(u_i) = y_i$. Then $V_i + W_i$ is a Jacobi field having values y_i at u_i, y_{i+1} at u_{i+1}, which shows existence of $Y|_{[u_i, u_{i+1}]}$. Moreover, if Y is given, $Y|_{[u_i, u_{i+1}]} - W_i = V_i|_{[u_i, u_{i+1}]}$ by the uniqueness of V_i, which shows uniqueness of $Y|_{[u_i, u_{i+1}]}$.

The linearity and one-to-one-ness of G follow easily from the uniqueness of Y and the linearity of the Jacobi equation.

We denote the range of G by $\mathscr{K}$. Thus $\dim \mathscr{K} = (d - 1) n$.

By evaluation of $V \in \mathscr{L}$ at $u_1, \ldots, u_n$ we get a linear transformation $E : \mathscr{L} \to \Sigma \tau_*(u_i)^{\perp}$, $E(V) = (V(u_1), \ldots, V(u_n))$. The composition

$F = G \circ E : \mathscr{L} \to \mathscr{K}$ is the infinitesimal version of the deformation $\phi : \sigma \to \gamma$, so we call F an *infinitesimal deformation*. Corresponding to the fact that ϕ is length-nonincreasing is the fact that F does not increase the index form I, which is proved below.

Problem 13. If Q is a broken C^∞ rectangle with the transverse curves τ^{u_i} lying in T_i, then we get another rectangle $\phi(Q)$ by applying ϕ to the longitudinal curves of Q. If V is the vector field associated to Q, show that $F(V)$ is the field associated to $\phi(Q)$.

Lemma 4. If Y and Z are Jacobi fields, then $\langle Y, Z' \rangle - \langle Y', Z \rangle$ is constant. If Y and Z are N-Jacobi fields the constant is 0.

Proof.
$$(\langle Y, Z' \rangle - \langle Y', Z \rangle)' = \langle Y, Z'' \rangle - \langle Y'', Z \rangle$$
$$= -\langle Y, R_{\tau_* Z} \tau_* \rangle + \langle R_{\tau_* Y} \tau_* , Z \rangle$$
$$= 0, \qquad \text{by a curvature identity.}$$

If Y and Z are N-Jacobi fields, then

(a) $\qquad\qquad\qquad \langle S_{\tau_*(0)} Y(0) - Y'(0), Z(0) \rangle = 0,$

(b) $\qquad\qquad\qquad \langle S_{\tau_*(0)} Z(0) - Z'(0), Y(0) \rangle = 0,$

so subtracting (a) and (b) and using the symmetry of $S_{\tau_*(0)}$ gives

$$-\langle Y'(0), Z(0) \rangle + \langle Z'(0), Y(0) \rangle = 0.$$

Problem 14. Let $\mathscr{J}$ be a subspace of the Jacobi fields along τ having dimension $d - 1$ and such that for every $Y, Z \in \mathscr{K}$,

$$\langle Y, Z' \rangle - \langle Y', Z \rangle = 0.$$

Show that there is a submanifold N normal to τ at $\tau(0)$ such that $\mathscr{J}$ is the space of N-Jacobi fields.

Theorem 4. *The basic inequality.* Suppose there is no focal point of N on $\tau((0, b])$. For $V \in \mathscr{L}$ there is a unique N-Jacobi field Y such that $Y(b) = V(b)$, by lemma 3. Then $I(V) \geqslant I(Y)$ and equality occurs if and only if $V = Y$.

Proof. Let $Y_1, ..., Y_{d-1}$ be a basis of N-Jacobi fields. Then there is a unique expression $V = \sum_{i=1}^{d-1} f_i Y_i$ valid on $(0, b]$, where the f_i are continuous broken C^∞ functions. We leave as an exercise the fact that such an expression exists and is unique on $[0, b]$ as well. (See problem 15.)

At points where V' exists we let $W = \Sigma_i f_i' Y_i$, and $Z = \Sigma_i f_i Y_i'$, so that $V' = W + Z$. Then

(a) $$\left\langle V, \sum_j f_j' Y_j' \right\rangle = \sum_{i,j} f_i f_j' \langle Y_i, Y_j' \rangle$$

$$= \sum_{i,j} f_i f_j' \langle Y_i', Y_j \rangle \qquad \text{(by lemma 4)},$$

$$= \langle W, Z \rangle.$$

(b) $\langle V, Z \rangle' = \langle V', Z \rangle + \langle V, Z' \rangle$

$$= \langle W, Z \rangle + \langle Z, Z \rangle + \left\langle V, \sum_i f_i' Y_i' \right\rangle + \left\langle V, \sum_i f_i Y_i'' \right\rangle$$

$$= \langle W, Z \rangle + \langle Z, Z \rangle + \langle W, Z \rangle - \left\langle V, \sum_i f_i R_{\tau_* Y_i} \tau_* \right\rangle$$

$$= 2 \langle W, Z \rangle + \langle Z, Z \rangle - \langle V, R_{\tau_* V} \tau_* \rangle.$$

The integrand in $I(V)$ is thus

(c) $\langle V', V' \rangle - \langle R_{\tau_* V} \tau_*, V \rangle = \langle W + Z, W + Z \rangle - \langle R_{\tau_* V} \tau_*, V \rangle$

$$= \langle W, W \rangle + \langle V, Z \rangle' \qquad \text{[by (b)]}.$$

(d) $I(V) = \langle S_{\tau_*(0)} V(0), V(0) \rangle - \langle S_{\tau_*(b)} V(b), V(b) \rangle$

$$+ \int_0^b (\langle W, W \rangle + \langle V, Z \rangle') \, du$$

$$= \langle S_{\tau_*(0)} V(0), V(0) \rangle - \langle S_{\tau_*(b)} V(b), V(b) \rangle$$

$$+ \langle V(b), Z(b) \rangle - \langle V(0), Z(0) \rangle + \int_0^b \langle W, W \rangle \, du.$$

However,

$$S_{\tau_*(0)} V(0) - Z(0) = \sum_i f_i(0)(S_{\tau_*(0)} Y_i(0) - Y_i'(0)) \perp V(0),$$

so

(e) $$I(V) = \langle Z(b) - S_{\tau_*(b)} V(b), V(b) \rangle + \int_0^b \langle W, W \rangle \, du.$$

The N-Jacobi field which coincides with V at b is $Y = \Sigma_i\, c_i Y_i$, where $c_i = f_i(b)$. Letting $W_1 = \Sigma_i\, c_i' Y_i = 0$, $Z_1 = \Sigma_i\, c_i Y_i'$, the same computation (a)-(e) gives, since $Z_1(b) = Z(b)$, $Y(b) = V(b)$,

(f) $$I(Y) = \langle Z(b) - S_{\tau_*(b)} V(b),\, V(b)\rangle.$$

Hence $I(V) - I(Y) = \int_0^b \langle W, W\rangle\, du \geqslant 0$. Equality occurs if and only if $W = 0$, which in turn is equivalent to $f_i' = 0$ for all i, f_i are constant for all i, and, finally, $Y = V$.

Corollaries 1 and 2, which follow, have the same hypothesis on N and $\tau((0, b])$, but not corollary 3.

Corollary 1. If $V(b) = 0$, then $I(V) \geqslant 0$, and equality occurs if and only if $V = 0$.

Corollary 2. Let $Y \in \mathscr{L}$. Then Y is an N-Jacobi field if and only if $I(V, Y) = 0$ for all $V \in \mathscr{L}$ such that $V(b) = 0$.

Proof. By polarization, (e) in the above proof gives, for V_1, $V_2 \in \mathscr{L}$, $V_1' = Z_1 + W_1$, $V_2' = Z_2 + W_2$, etc.,

$$I(V_1, V_2) = \langle Z_1(b) - S_{\tau_*(b)} V_1(b), V_2(b)\rangle + \int_0^b \langle W_1, W_2\rangle\, du.$$

If Y is an N-Jacobi field and $V(b) = 0$, letting $V_1 = Y$, $V_2 = V$ gives $W_1 = 0$, $V_2(b) = 0$, so $I(V_1, V_2) = I(Y, V) = 0$.

Conversely, suppose $I(V, Y) = 0$ for all $V \in \mathscr{L}$ such that $V(b) = 0$. Let Y_1 be the N-Jacobi field such that $Y(b) = Y_1(b)$. Then $I(Y, Y - Y_1) = 0$ by hypothesis, $I(-Y_1, Y - Y_1) = 0$ as just proved, so

$$I(Y - Y_1) = I(Y, Y - Y_1) + I(-Y_1, Y - Y_1) = 0.$$

Hence by corollary 1, $Y - Y_1 = 0$.

Corollary 3. Suppose that there is no conjugate point of $\tau(0)$ on $\tau((0, b])$. For $V \in \mathscr{L}$ there is a unique Jacobi field Y such that $Y(0) = V(0)$, $Y(b) = V(b)$, by lemma 3. Then $I(V) \geqslant I(Y)$ and equality occurs if and only if $V = Y$.

Proof. Let T_1 and T_2 be $(d-1)$-dimensional transverse manifolds at $\tau(0)$ and $\tau(b)$, respectively, such that their second fundamental forms vanish, and $I_1 = I(T_1, T_2)$ the associated index form. Then for $V_1, V_2 \in \mathscr{L}$

$$I_1(V_1, V_2) - I(V_1, V_2) = -\langle S_{\tau_{*(0)}} V_1(0), V_2(0) \rangle + \langle S_{\tau_{*(b)}} V_1(b), V_2(b) \rangle,$$

and hence $I_1(V) - I(V) = I_1(Y) - I(Y)$, so it suffices to prove the result for I_1 instead of I.

Let $I_2 = I(\tau(0), T_2)$, $I_3 = I(T_1, \tau(b))$, so that I_1, I_2, and I_3 are the same except for their domains.

From lemma 3, $Y = Y_1 + Y_2$, where Y_1 is the $\tau(0)$-Jacobi field such that $Y_2(0) = V(0)$. Since $V - Y$ is 0 at both $\tau(0)$ and $\tau(b)$, $I_1(V - Y, Y_1) = I_2(V - Y, Y_1) = 0$ and $I_1(V - Y, Y_2) = I_3(V - Y, Y_2) = 0$, by corollary 2 applied to I_2 and I_3, respectively. Adding these two equations gives

$$I_1(V - Y, Y_1 + Y_2) = I_1(V - Y, Y) = 0, \qquad \text{or,} \qquad I_1(V, Y) = I_1(Y, Y).$$

But by corollary 1,

$$0 \leqslant I_2(V - Y, V - Y) = I_1(V, V) - 2I_1(V, Y) + I_1(Y, Y)$$

$$= I_1(V, V) - I_1(Y, Y),$$

and equality occurs if and only if $V - Y = 0$.

Problem 15. If $V_1, ..., V_d$ are C^∞ vector fields along τ and are independent except at $\tau(0)$, and V is a broken C^∞ vector field along τ, with $V = \Sigma f_i V_i$ the expression for V in terms of the V_i valid on $(0, b]$, show by an example that the f_i need not have continuous extensions to $[0, b]$ even if $V(0) = \Sigma a_i V_i(0)$. However, if the V_i are Jacobi fields then there are unique continuous extensions of the f_i.

Theorem 5. An infinitesimal deformation F is I-nonincreasing; that is, $I(V) \geqslant I(F(V))$. Moreover, equality is obtained only if $V = F(V)$.

Problem 16. Prove theorem 5 by applying theorem 4 and corollary 3.

Corollary 1. (For $N =$ point this is due to Jacobi.) τ does not minimize distance to N beyond the first focal point.

Proof. Suppose $\tau(c)$ is a focal point of N, $c \in (0, b)$, and let Y be a nonzero N-Jacobi field which vanishes at c. Let $V \in \mathscr{L}$ be given by

$V \mid_{[0,c]} = Y \mid_{[0,c]}$, $V \mid_{[c,b]} = 0$. Let $I = I(N, \tau(b))$, $I_1 = I(N, \tau(c))$, $I_2 = I(\tau(c), \tau(b))$. Then $I(V) = I_1(V) + I_2(V) = I_1(Y) + I_2(0) = 0$ by corollary 2 to theorem 4.

Now choose intermediate manifolds for an infinitesimal deformation F so that c is not one of the u_i. Then $F(V)$ is smooth at c, while V has a break at c, hence $F(V) \neq V$ and $I(F(V)) < I(V) = 0$. By theorem 3 there are shorter curves from $\tau(b)$ to N.

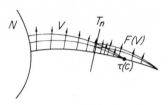

FIG. 40.

Corollary 2. The first focal point of N is that point $\tau(c)$ such that $\tau([0, b])$ fails to minimize arc length to N for $b > c$, but $\tau([0, b'])$ minimizes arc length to N among curves in a neighborhood of $\tau([0, b'])$ for $b' < c$.

The *augmented index* of a quadratic form is the dimension of a maximal subspace on which the form is negative semidefinite.

Corollary 3. The index and augmented index of I are the same as the index and augmented index of $I \mid_{\mathscr{K}}$, hence are finite.

Proof. Since it is obvious that the index of I is not less than that of of $I \mid_{\mathscr{K}}$ it suffices to prove the inequality the other way, and similarly for the augmented index.

Let $\mathscr{H}$ be a subspace of $\mathscr{L}$ on which I is negative semidefinite. Then $F \mid_{\mathscr{H}}$ is an isomorphism. For if $V \in \mathscr{H}$ is in the kernel of F, then $F(V) = 0, I(F(V)) = 0 \leqslant I(V) = 0$, so from equality, $V = F(V) = 0$. Since I is negative semidefinite on $F(\mathscr{H})$ this proves the desired inequality for the augmented index. For the index the argument is the same except for using "definite" in place of "semidefinite."

Now let N have focal points at distances c_i along τ, $i = 1, ..., h_-$, where $c_i \leqslant c_{i+1} < b$ and each c_i is included a number of times equal to the multiplicity of $\tau(c_i)$ as a focal point; we do not exclude the possibility that $\{c_i\}$ is a proper subset of the set of *focal values* in $(0, b)$, and we do not know *a priori* that the number of focal points is finite. Thus there are fields $Y_i \in \mathscr{L}$, $i = 1, ..., h_-$, such that $Y_i |_{[0, c_i]}$ is an N-Jacobi field, $Y_i |_{[c_i, b]} = 0$, and for each $c \in (0, b)$, $\{Y_i \mid c = c_i\}$ are independent. Let $\mathscr{H}_-$ be the linear space spanned by these Y_i, and let $\mathscr{H} = \mathscr{H}_- + \mathscr{H}_0$, where $\mathscr{H}_0$ is the null space of I.

Problem 17. Prove that the Y_i are independent, so the dimension of $\mathscr{H}_-$ is h_-. Moreover, the sum $\mathscr{H}_- + \mathscr{H}_0$ is direct.

Lemma 5. The restriction of I to $\mathscr{H}$ is identically 0.

Proof. Since $I(\mathscr{H}, \mathscr{H}_0) = 0$ it suffices to show $I(\mathscr{H}_-, \mathscr{H}_-) = 0$, or that $I(Y_i, Y_j) = 0$ for every i, j. We may assume $c_i \leqslant c_j$. Let $I_1 = I(N, \tau(c_j))$, $I_2 = I(\tau(c_j), P)$. Then

$$I(Y_i, Y_j) = I_1(Y_i, Y_j) + I_2(Y_i, Y_j) = I_1(Y_i, Y_j) = 0$$

since Y_j is in the null space of I_1 by theorem 1.

Let $a(I)$, $i(I)$, and $n(I)$ denote the augmented index, the index, and the nullity of I, respectively. It is well known that on a finite-dimensional space the sum of the index and nullity of a quadratic form equals the augmented index. Because we have been able to reduce I to a finite-dimensional subspace $\mathscr{H}$ without altering the index and augmented index, this result is also true for I, so $a(I) = i(I) + n(I)$.

Henceforth, when we speak of the *number* of focal points we shall mean the sum of their orders.

Theorem 6. The number of focal points of N on $\tau((0, b))$ is finite and is a lower bound for $i(I)$.

Proof. By problem 17 and lemma 5 and the fact that $a(I)$ is finite, $h_- + n(I) = \dim \mathscr{H} \leqslant a(I) = i(I) + n(I)$, so $h_- \leqslant i(I)$.

11.5 The Morse Index Theorem [57, 83]

In this section we restrict to the case where $P = \tau(b)$. Then the Morse index theorem says that the inequality of theorem 6 is an equation.

Theorem 7. Let $I = I(N, \tau(b))$. Then $a(I)$ is the number of focal points of N on $\tau((0, b])$.

Proof. The idea is to examine $I_t = I(N, \tau(t))$ and the integer-valued functions $a(t) = a(I_t)$, $i(t) = i(I_t)$, and $n(t) = n(I_t)$ as t passes from 0 to b.

Because the domain of I_t may be regarded as an increasing function of t, a, and i are nondecreasing; for small t, I_t is positive-definite, so a and i are initially 0.

We have already noted that $a = i + n$, and that $n(t) = 0$ except for a finite number of t, where $n(t)$ is the order of the focal point. Thus a must have jumps which at least add up to the sum of the $n(t)$ (theorem 6).

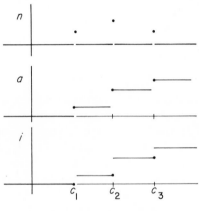

Fig. 41.

On the other hand, because I_t is continuous in t, in order for some of the positive definite part to go into the negative definite part it must pass through the null part. This causes the jumps of a and i to occur separately, on opposite sides of the focal values, and with magnitude $n(t)$. As a value t is entered from the left the jump of a is $n(t)$, i is continuous from the left; formally, $a(t) = a(t^-) + n(t)$, $i(t) = i(t^-)$. When we pass on from t a is continuous from the right, n drops to 0, and i must take up the slack in the equation $a = i + n$; $a(t^+) = a(t)$, $i(t^+) = i(t) + n(t)$. So what we need to establish in detail is, for every t, $a(t^+) = a(t)$ and $i(t^-) = i(t)$; the theorem then follows by a simple computation.

Let F_t be an infinitesimal deformation for I_t having intermediate manifolds at $\tau(u_i)$, $i = 1, ..., n$. Then the same intermediate manifolds will serve to give an infinitesimal deformation F_u for I_u when $u \in U$, a neighborhood of t. This infinitesimal deformation is given by the evaluation map E and an isomorphism $G_u : \mathscr{G} \to \mathscr{K}_u$, where $\mathscr{G} = \Sigma \perp(u_i)$ is the direct sum of the tangent spaces of the intermediate manifolds. The map $H_u = G_u G_t^{-1} : \mathscr{K}_t \to \mathscr{K}_u$ is an isomorphism, and it alters only that part of $Y \in \mathscr{K}_t$ defined on $(u_n, t]$; the part defined on $[0, u_n]$ is unchanged. The value of I_t on Y is a sum of terms of which only one, $-\langle Y'(u_n{}^+), Y(u_n)\rangle$, is altered by H_u, and even here $Y(u_n)$ remains fixed.

Let $Y_u = H_u(Y)$. In order to prove continuity of $I_u(Y_u)$ as a function of u it is sufficient to prove that $Y_u'(u_n{}^+)$ is continuous in u. By its definition Y_u, restricted to $[u_n, u]$, is the Jacobi field which satisfies $Y_u(u) = 0$ and $Y_u(u_n) = Y(u_n) = y$. For y, $z \in \perp(u_n)$ let $Y_{y,z}$ be the unique Jacobi field such that $Y_{y,z}(u_n) = y$ and $Y'_{y,z}(u_n) = z$; the map $(y, z) \to Y_{y,z}$ is linear in y and z, and so is evaluation at u, so we may write $Y_{y,z}(u) = A_u(y) + B_u(z)$, where

$$A_u, B_u : \perp(u_n) \to \perp(u)$$

are linear transformations. A_u and B_u are continuous in u since $Y_{y,z}$ is continuous. However, there is a unique solution for z of $A_u(y) + B_u(z) = 0$, so that B_u^{-1} exists and is continuous. Thus the solution $Y_u'(u_n{}^+) = z = -B_u^{-1}A_u(y)$ is continuous in u, and finally so is $I_u(Y_u)$.

Now let $\mathscr{K}^+$ be a maximal subspace of $\mathscr{K}_t$ on which I_t is positive definite and let S^+ be the unit sphere of $\mathscr{K}^+$ with respect to I_t. Then $a(t) = \text{codim } \mathscr{K}^+$. Then $f : (u, Y) \to I_u(Y_u)$ is a continuous function on $U \times S^+$ which is positive on $\{t\} \times S^+$. Since S^+ is compact, there is a neighborhood U' of t such that f is positive on $U' \times S^+$ (see lemma 6 below). Thus for $u \in U'$, I_u is positive-definite on the space $H_u(\mathscr{K}^+)$ spanned by $H_u(S^+)$, and $a(u) \leqslant a(t)$. But a is nondecreasing, so $a(t^+) = a(t)$.

The same argument applies to a maximal negative definite subspace of I_t on $\mathscr{K}_t$ to show that $i(u) \geqslant i(t)$ for u in some neighborhood of t. Thus $i(t^-) = i(t)$.

Lemma 6. Let f be a continuous real-valued function on a fibre bundle B with compact fibre F and base M; denote by F_p the fibre over $p \in M$. Define $f_m, f_M : M \to R$ by

$$f_m(p) = \min\{f(b) \mid b \in F_p\}$$
$$f_M(p) = \max\{f(b) \mid b \in F_p\}.$$

Then f_m and f_M are continuous.

Proof. Since this is a local theorem, we may assume that $B = F \times M$. For $\epsilon > 0$ we can find a finite covering $W_i = V_i \times W$ of $F \times \{p\}$, W a neighborhood of p, such that $|f(b) - f(b')| < \epsilon$ for every b, $b' \in W_i$. Then for $x \in F$ and $p' \in W$, $|f(x, p') - f(x, p)| < \epsilon$. Now if $b_1 = (x_1, p)$ is a point where $f(b_1) = f_M(p)$ and $b_2 = (x_2, p')$ is a similar point for $f_M(p')$, then

$$f_M(p') - \epsilon = f(x_2, p') - \epsilon < f(x_2, p) \leqslant f(x_1, p)$$

$$= f_M(p) < f(x_1, p') + \epsilon \leqslant f(x_2, p') + \epsilon = f_M(p') + \epsilon,$$

so $|f_M(p') - f_M(p)| < \epsilon$, which concludes the proof for f_M. The proof for f_m follows by considering $-f$.

Problem 18. Let J be the inner product on $\mathscr{L}$, the domain of $I(N, P)$, given by $\bar{J}(Y, Z) = \int_0^b \langle Y, Z \rangle \, du$. Also let $J(Y) = \bar{J}(Y, Y)$. A *characteristic value* of I with respect to J is a number λ such that $n(I - \lambda J) \neq 0$; a *characteristic vector* of I belonging to λ is a vector field in the null space of $I - \lambda J$. Show that:

(a) For λ sufficiently negative $I - \lambda J$ is positive-definite.

(b) Y is a characteristic vector of I if and only if Y satisfies the end conditions at N and P ($S_{\tau_*} Y - Y'$ is perpendicular to N, P) and is a solution of the second order equation $Y'' + R_{\tau_* Y} \tau_* + \lambda Y = 0$.

(c) The augmented index of $I - \lambda J$ is finite and is equal to the number of independent characteristic vectors belonging to values $\leqslant \lambda$.

(d) Let $P = \tau(b)$ and let J_t be the restriction of J to the domain of I_t. Number the characteristic values of I_t, counting multiplicities, starting with the least, so as to obtain a sequence of functions $\{\lambda_i\}$ of t. Then the λ_i are continuous and nonincreasing on $(0, b]$ and $n(I_t)$ is the number of i such that $\lambda_i(t) = 0$.

Problem 19. With each $x \in \underline{\perp}(N)$, x not in the 0 cross section, associate the index form $I_x = I(N, \exp x)$ at the geodesic $\exp ux \mid_{[0,1]}$. Show that:

(a) If $n(I_x) = 0$, there is a neighborhood U of x such that for every $x' \in U$, $i(I_{x'}) = i(I_x)$.

(b) If M is complete and N is closed in M, the set $\{m \mid n(I_x) = 0$ for every nonzero $x \in \exp^{-1} m\}$ is everywhere dense in M (see problem 1.11).

Those interested in studying Morse theory and its applications are referred to [*37, 38, 54, 57, 71, 81*].

11.6 The Minimum Locus [*46, 60, 61, 93*]

A *minimal segment* is a geodesic segment which minimizes arc length between its ends. A *minimum point m of p along a geodesic* γ is a point on γ such that the segment of γ from p to m is minimal but no larger segment from p is minimal. The set of all minimum points of p is called the *minimum* (or *cut*) *locus* of p.

It follows immediately from the fact that geodesics do not minimize arc length beyond the first conjugate point that if m is the first conjugate point of p along γ, then there is a minimum point of p along γ which is not beyond m.

A geodesic ray from p has at most one minimum point of p, but there may be none.

If $t \in M_p$ is such that $\exp_p t$ is the minimum point of p along the geodesic $\exp_p ut$, then we also say that t is a *minimum point of p in* M_p .

Theorem 8. (a) If m is not a minimum point of p, then there is at most one minimal segment γ from p to m. If $\{\sigma_i\}$ is a sequence of curves from p to m such that $\lim |\sigma_i| = \rho(p, m)$, then σ_i converges to γ (if properly parametrized).

(b) If there is a minimal segment from p to m on which m is a conjugate point of p, then m is a minimum point of p.

(c) If M is complete the converse is true: if m is a minimum point of p, then either there are two minimal segments or m is a conjugate point of p along the unique one.

Proof. (a) The first statement follows from the second, because if γ and σ were both minimal segments letting $\sigma_i = \sigma$ for all i would give $\lim \sigma_i = \sigma = \gamma$.

We suppose that γ and each σ_i are parametrized by arc length. By compactness of a normal coordinate ball of some radius ϵ at m there will be convergent subsequences of $\{\sigma_i(L_i - \epsilon)\}$, where $L_i = |\sigma_i|$. If one of these subsequences did not converge to $\gamma(L - \epsilon)$, where $L = \rho(p, m)$, then the corresponding curves would eventually all form a corner with the extension of γ beyond m. By cutting across this corner we would get shorter curves from p to $\gamma(L + \epsilon)$ than $L + \epsilon$, contradicting the minimality of γ beyond m. By using triangle inequalities and local minimizing of geodesics only, the argument can be made precise, and shows even more, that σ_i converges to γ within a regular neighborhood of m. By covering γ with a finite number of regular neighborhoods the convergence can be shown inductively down γ from m to p. By "regular" we mean that every two points in the neighborhood can be joined by a unique minimal segment. The details are left as an exercise.

(b) follows immediately from the fact that geodesics do not minimize arc length beyond the first conjugate point.

(c) Suppose M is complete, and m is a minimum point of p such that there is a unique minimal segment γ. Then we must show that m is a conjugate point along γ.

Let $L = \rho(p, m)$ and let σ_i be a minimal segment from p to $m_i = \gamma(L + 1/i)$, parametrized by arc length, and with length L_i. Then $\{\sigma_{i*}(0)\}$ must have limit points, but the geodesic in the direction of such a limit point would give a minimal segment from p to m; hence, $\lim \sigma_{i*}(0) = \gamma_*(0)$. But then

$$\exp_p(L_i \sigma_{i*}(0)) = m_i = \exp_p \left(L + \frac{1}{i} \right) \gamma_*(0)$$

and

$$\lim L_i \sigma_{i*}(0) = \lim \left(L + \frac{1}{i} \right) \gamma_*(0) = L \gamma_*(0),$$

so $\exp_p$ is not one-to-one in a neighborhood of $L\gamma_*(0)$. Thus $d \exp_p$ is singular there and m is a conjugate point.

Problem 20. In (a) show that the convergence of σ_i to γ is uniform.

Problem 21. Show that the relation "is a minimum point of" is symmetric. Thus if γ is a minimal segment from p to minimum point m there are shorter curves than γ from m to the points on the extension of γ from p away from m.

Problem 22. Assume M is connected. Define m to be *between* p and q if all three points are distinct and $\rho(p, m) + \rho(m, q) = \rho(p, q)$; this relation is denoted by $[p, m, q]$. Show:

(a) If $[p, m, q]$, $[p, n, q]$, and $[p, m, n]$, then $[m, n, q]$ and a minimal segment γ from m to n is unique. In case γ exists let σ be the largest (open) geodesic extension of γ which does not include p or q. Then every point on σ is between p and q. Hence $| \sigma | \leqslant \rho(p, q)$ and σ has no minimum points of any of its points.

(b) If M is not complete then for every $p \in M$ there is a geodesic ray from p which has no minimum point of p on it.

(c) If M is complete then $[p, m, q]$ if and only if m is in an open minimal segment with ends p and q.

Let $Q(M) = \{t \in T(M) \mid t$ is a unit vector and $c_t t$ is a minimum point of $\pi't$ for some $c_t > 0\}$. Define $f : Q(M) \to M$ by $f(t) = \exp c_t t$; that is, f assigns to a tangent t the minimum point of the base point of t along the geodesic in the direction of t, if there is one.

Theorem 9. f is continuous.

Proof. For $\{t_i\}$, a convergent sequence in $Q(M)$, we wish to show that $\lim f(t_i) = f(\lim t_i)$. Let $\gamma_i = \exp ut_i$, $p_i = \gamma_i(0)$, $m_i = f(t_i) = \gamma_i(c_i)$, $t = \lim t_i$, $\gamma = \exp ut$, $p = \gamma(0)$, and $m = f(t) = \gamma(c)$. It will be sufficient to show $\lim c_i = c$, for then by continuity of exp,

$$\lim m_i = \lim \exp c_i t_i = \exp \lim c_i t_i = \exp (\lim c_i \lim t_i) = \exp ct = m.$$

Suppose $\lim \sup c_i > c$. Then there are $\epsilon > 0$, a subsequence $\{d_i\}$ of $\{c_i\}$, and k such that for $i > k$, $d_i > c + \epsilon$ and $\gamma(c + \epsilon)$ is defined. Then for the corresponding subsequences $\{\sigma_i\}$ of $\{\gamma_i\}$, $\{n_i\}$ of $\{m_i\}$, and $\{q_i\}$ of $\{p_i\}$, σ_i is a minimal segment from q_i to $\sigma_i(c + \epsilon)$, so

$$c + \epsilon = \lim \rho(q_i, \sigma_i(c + \epsilon))$$
$$= \rho (\lim q_i, \lim \sigma_i(c + \epsilon))$$
$$= \rho(p, \gamma(c + \epsilon)),$$

which contradicts the fact that γ does not minimize arc length beyond $\gamma(c) = m$.

To show $\lim \inf c_i = c$ we consider convergent subsequences; thus we may suppose that $\lim c_i = c' < c$ and reach a contradiction. Let $\epsilon = (c - c')/2 > 0$. Then $q_i = \gamma_i(c_i + \epsilon)$ is beyond the minimum point of p_i on γ_i, so there is a shorter curve τ_i from p_i to q_i; by adding to τ_i short segments from p to p_i and from q_i to $q = \gamma(c' + \epsilon)$ we get a curve σ_i from p to q such that $\lim | \sigma_i | = \rho(p, q)$. By theorem 8(a), σ_i converges to γ, and hence τ_i converges to γ also.

Let $E = \pi' \times \exp : T(M) \to M \times M$. Then E is nonsingular on the compact set $\{ut \mid 0 \leqslant u \leqslant c' + \epsilon\}$, so by problem 1.12 there is a neighborhood U on which E is a diffeomorphism. Let $V = E(U)$, so V is a neighborhood of $E(ut) = (p, \gamma(u))$. For sufficiently large i both $(p_i, \gamma_i(u))$ and $(p_i, \tau_i(u))$ will be in V. However, $E^{-1}(p_i, q_i)$ must have length $c_i + \epsilon$ because $q_i = \exp_{p_i}(c_i + \epsilon) t_i$ and on the other hand cannot have length any greater than the integral of the radial lengths of τ_i, which is less than $c_i + \epsilon$. QED

Corollary 1. The distance from a fixed point p to its minimum point in the direction $t \in M_p$ is a continuous function of t where defined.

(This is immediate from the fact that the distance function is continuous.)

Assume now that M is connected.

Corollary 2. A Riemannian manifold M is compact if and only if for some point p there is a minimum point in every direction from p.

Proof. If M is compact then M is complete and bounded. Thus every geodesic ray can be extended indefinitely but cannot minimize arc length beyond the bound on M.

Conversely, if p is a point such that every geodesic ray from p has a minimum point, then the function $g : S \to R$, where S is the unit sphere in M_p, $g(t) = \rho(p, \exp c_t t)$, is continuous by corollary 1. Thus the set

$$B = \{t \in M_p \mid \| t \| \leqslant g(t/\| t \|) \quad \text{or} \quad t = 0\}$$

is closed and bounded in M_p, hence compact. But then $M = \exp B$ is compact. (See problem 22(b).)

Corollary 3. If every geodesic ray from a point p has a conjugate point of p, then M is compact.

Corollary 4. If a complete M has a covering space which is not compact, then for every $p \in M$ there is a geodesic ray on which there are no points conjugate to p.

Proof. We may assume the covering is a Riemannian covering. Hence, the projection preserves geodesics and so also conjugate points. By corollary 3 there will be a geodesic from a point above p which has no conjugate points and then the projection of this geodesic will be without conjugate points.

Corollary 5. If M is complete, then the distance to the minimum locus is a continuous function of the point.

Proof. Let $f(p) = $ distance from p to its minimum locus,

$$g(p) = \begin{cases} e^{-f(p)} & \text{if } p \text{ has a minimum point} \\ 0 & \text{otherwise.} \end{cases}$$

It suffices to show that g is continuous. Let $h(p, t)$ be the distance from p to the minimum point of p on the geodesic in the direction of $t \in M_p$. Then h is continuous on $Q(M)$ and

$$\bar{g}(p, t) = \begin{cases} e^{-h(p,t)} & \text{if } h(p, t) \text{ is defined} \\ 0 & \text{otherwise} \end{cases}$$

is continuous on all of the unit tangent bundle, since if $(p, t) \in T(M)$ is a limit point of $Q(M)$, it follows from the first part of the proof of theorem 9 that $\lim_{x \to (p,t)} h(x) = +\infty$.

Since it is clear that g is the maximum-on-fibres of $\bar{g}$ the result now follows from lemma 6.

Remark. Corollary 1 shows that the complement of the minimum locus of a point is topologically a cell when M is complete, and in fact the exponential map gives the homeomorphism. Hence, much of the topological interest of a manifold lies in its minimum locus.

Problem 23. Extend the results of this section, except for theorem 9 to the case where p is replaced by a closed or a compact submanifold N, using which ever the result requires. In particular, show by examples that the extensions of corollaries 2, 3, and 4 require N to be compact.

11.7 Closed Geodesics

A *closed geodesic* is a geodesic segment for which the initial and final points coincide; a *smooth* closed geodesic is one for which the initial and final tangents coincide.

It is intuitively clear that a homotopy class of loops based at p should have a closed geodesic for its minimal-length representative, provided the manifold is complete. Indeed, such a class is represented in the simply connected covering by the class of all curves from p_0 to p_1, where p_0 and p_1 are in the fibre over p. A minimal segment from

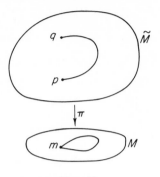

FIG. 42.

p_0 to p_1 projects to a closed geodesic in the class, and is clearly a minimal-length curve in the class. To make the identity in the fundamental group nonexceptional we must consider the constant curve as being a closed geodesic.

To get smooth closed geodesics we consider free homotopy classes of loops. These classes are in one-to-one correspondence with the conjugacy classes of the fundamental group, because, roughly, the isomorphism between the fundamental groups based at different points, which free homotopy ignores, is determined only up to an inner automorphism, or conjugation. It is not necessarily true that such a class will have a minimal-length member, even if M is complete, because tightening up a loop might force it to go to "infinity." For example, this is the case for the surface obtained by revolving the curve $xz = 1$ around the z-axis.

However, if M is compact there is a minimal member in each free homotopy class. For such a class A will be represented by the lifts of its members to the simply connected covering $\tilde{M}$. Let

$$b = \inf\{\rho(p, q) \mid \text{if } \sigma \text{ is a curve from } p \text{ to } q \text{ then } \pi \circ \sigma \in A\},$$

where $\pi : \tilde{M} \to M$ is the covering projection. By taking a sequence $\{(p_i, q_i)\}$ such that $\lim \rho(p_i, q_i) = b$, and a sequence of minimal segments γ_i from p_i to q_i, a convergent subsequence of the closed geodesics $\pi \circ \gamma_i$ can be extracted, using the compactness of M. The limit will be a minimal closed geodesic in A, and it will be smooth because it is a minimal member of the homotopy class based at each of its points.

Theorem 10. If M is a compact Riemannian manifold, then every free homotopy class of loops has a minimal-length member which is a smooth closed geodesic.

Problem 24. Let σ be a loop in free homotopy class A. Then σ may be approximated uniformly by a broken geodesic loop $\gamma_0 \in A$. Let M be compact. Then we may assume that the breaks of γ_0 are closer together than the distance from any point to its minimum locus. Construct a sequence of broken geodesics inductively by letting γ_i be the broken geodesic loop which has as its segments the minimal segments between the midpoints of the segments of γ_{i-1}. Show that $\gamma_i \in A$ and that a subsequence converges to a smooth closed geodesic in A.

Theorem 11 (Synge's theorem [87]). If M is compact, orientable, even-dimensional, and has positive sectional curvatures, then M is simply connected.

Proof. The idea is to use second variation to show that a nontrivial smooth closed geodesic cannot be minimal, from which it follows by theorem 10 that there is only the trivial homotopy class of loops, M is simply connected.

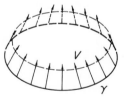

FIG. 43

Let γ be a smooth closed geodesic. Then parallel translation once around γ is an orthogonal transformation T of the odd-dimensional normal space to γ. Since M is orientable, T has determinant 1, and hence must have at least one characteristic value equal to 1. (The characteristic values have absolute value 1 and the nonreal ones occur in conjugate pairs.) The characteristic vectors for the value 1 are fixed points of T, so there is a smooth parallel field V along γ. Letting N

be any transverse manifold, $I = I(N, N)$, the end terms of $I(V)$ cancel and $V' = 0$, so

$$I(V) = - \int_0^b K(\gamma_*, V) \langle V, V \rangle \, du.$$

Since curvature is positive this is negative and there are shorter nearby curves; γ cannot be minimal. QED

Problem 25. Let M be compact, even-dimensional, nonorientable, and have positive curvature. Show that the fundamental group of M is Z_2.

Problem 26. Let M be compact, odd-dimensional, and have positive sectional curvatures. Show that M is orientable.

Problem 27. Let M be a compact Kähler manifold which has positive holomorphic curvature. Show that M is simply connected. [*Hint*: if γ is a geodesic and J is the complex structure operator, then $J(\gamma_*)$ is parallel along γ and γ_*, $J(\gamma_*)$ span a holomorphic section.]

By using properties of the minimum locus it is sometimes possible to show the existence of closed geodesics with a method of Klingenberg [*45*].

Theorem 12. Let M be complete, $p \in M$ such that the minimum locus of p is nonempty, and let m be a point on the minimum locus of p which is closest to p. If m is not a conjugate point of p, then there is a unique closed geodesic with ends at p and passing through m such that both segments are minimal.

Proof. By theorem 8, if m is not a conjugate point, then there are at least two minimal segments from p to m. We show that there are exactly two and that they match smoothly at m. Let γ_1 and γ_2 be any two. If they do not match smoothly at m, then there is a geodesic σ starting at m which makes an acute angle with each of γ_1 and γ_2.

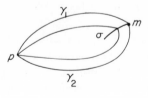

FIG. 44.

There will be minimal segments near to γ_1, from p to points on σ near m, and shorter than γ_1; similarly, there will be such minimal segments near to γ_2. Since γ_1 and γ_2 are distinct these minimal segments will also be distinct when the points on σ are sufficiently close to m. Then the points on σ will be minimum points of p, by theorem 8, but this contradicts the fact that m is the closest minimum point to p.

Corollary 1. Let M be compact and let (p, m) be a pair which realizes the minimum of the distances of points to their minimum locus. Then either p and m are conjugate to each other or there is a unique smooth closed geodesic through p and m such that both segments are minimal.

Corollary 2. Let M be compact, even-dimensional, orientable, with positive curvature, and let p, m be as in corollary 1. Then p and m are conjugate.

Proof. Assume p, m are not conjugate, so by corollary 1, there is a unique smooth closed minimal geodesic loop γ through p and m, $\gamma(0) = p$. Using Synge's trick, we have a one-parameter family of smooth loops γ_u such that $\gamma_0 = \gamma$ and $|\gamma_u| < |\gamma|$ for $u \neq 0$. Then the

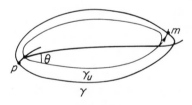

Fig. 45.

unique minimal segments from $\gamma_u(0)$ to the other points of γ_u form all possible angles with γ_u at $\gamma_u(0)$. Those which form a fixed angle θ have a convergent subsequence, as a function of u, to a minimal segment from p to a point m' on γ. By the uniqueness of m as the minimum point of p on γ, $m' = m$. This contradicts the fact that there can only be two minimal segments from p to m.

Remark. This corollary shows that under these conditions there is a point at which the conjugate locus and cut locus intersect. This can be used to derive lower bounds on diameter from upper bounds on curvature. Under the much stronger assumption that M is a simply connected Riemannian symmetric space the minimum and first conjugate locus coincide [28, 76].

11.8 Convex Neighborhoods [24; 33, p. 53; 93]

A set B in a Riemannian manifold M is *convex* if for every $m, n \in B$, there is a unique minimal segment from m to n and this segment is in B. The open ball $B(m, r_0)$ of radius r_0 about m is *locally convex* if each

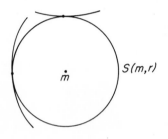

FIG. 46.

sphere $S(m, r)$ of radius $r < r_0$ about m satisfies the *convexity condition*: if γ is a geodesic tangent to $S(m, r)$ at $n = \gamma(0)$, then for sufficiently small u, $\rho(m, \gamma(u)) \geqslant r$. If $B(m, r_0)$ is locally convex then $\exp_m$ must be one-to-one on $B(0, r_0) \subset M_m$, for otherwise there would be in $B(m, r_0)$ points on the minimum locus of m; if γ were perpendicular to a geodesic τ from m at a point $\tau(r)$ beyond the minimum point, then $\rho(m, \gamma(u)) < r$ for small u, since $\rho(m, \tau(r)) < r$.

The relation between the concepts of convexity and local convexity is not as simple as it is in Euclidean spaces. For example, on a flat cylinder a normal coordinate ball with diameter greater than half the circumference of the cylinder will be locally convex but will not be convex because it will contain opposite points, which have two minimal segments. On the other hand, if the convexity condition fails for $S(m, r)$, then $B(m, r)$ is not convex. To show this let γ be a geodesic tangent to $S(m, r)$ at $n = \gamma(0)$ and having $p = \gamma(u)$ near n inside $S(m, r)$. Then a Jacobi field along γ which points outward at p and vanishes at n will have a corresponding rectangle having longitudinal geodesics, of which only γ is tangent to $S(m, r)$. Thus there will be

11.8. Convex Neighborhoods

segments which start near p, pass outside $S(m, r)$, and return in $S(m, r)$ at n. However, it might happen that $B(m, r')$ is convex some $r' > r$.

The example of a flat cylinder shows that the following is the best possible result of its kind.

Proposition 1. Let $B(m, 2r_0)$ be locally convex. Then every minimal segment between a pair in $B(m, r_0)$ is entirely within $B(m, r_0)$.

Proof. If p, $q \in B(m, r_0)$, then a minimal segment between them cannot go outside $B(m, 2r_0)$. If $\rho(m, \gamma)$ does not take on its maximum at an end of γ, then the least parameter value of γ for which $\rho(m, \gamma)$ is maximum would give a point of tangency of γ with a sphere $S(m, r)$, $r < 2r_0$, and one end of γ from that point of tangency would be inside $S(m, r)$, contradicting local convexity. Thus the maximum of $\rho(m, \gamma)$ occurs at an end, so all of γ lies inside $B(m, r_0)$.

If τ is a geodesic from m to $n = \tau(r)$, then $N = \exp_n(\tau_*(r)^\perp) \cap U$, where U is a neighborhood of n, is a submanifold containing all small geodesic segments tangent to $S(m, r)$ at n. Thus the index form $I = I(m, N)$ will determine largely whether $S(m, r)$ satisfies the convexity condition at n. If I has nonzero index, then there will be points on N which are closer to m than n is. Thus if the convexity condition on $S(m, r)$ is satisfied I will be positive-semidefinite; if I is positive-definite we need only the requirement that n be before the minimum point of m on τ to obtain the convexity condition at n for $S(m, r)$.

In turn, whether or not I is positive (semi-)definite is determined by the behavior of m-Jacobi fields along τ. For if V is an m-Jacobi field the end terms in $I(V)$ are 0, because the second fundamental form of N at n is 0, so $I(V) = \langle V'(r), V(r) \rangle$, by corollary 3 to lemma 2. We summarize the result as follows:

Proposition 2. Let $\mathcal{H}$ be the space of m-Jacobi fields along geodesic τ from m, parametrized by arc length. Let b_r be the quadratic form on $\mathcal{H}$ defined by $b_r(V) = \langle V'(r), V(r) \rangle$.

(a) If $B(m, r_0)$ is locally convex, then b_r is positive-semidefinite for $r \in (0, r_0)$.

(b) If $B(m, r_0)$ is a normal coordinate ball and b_r is positive-definite for all such τ and for all $r \in (0, r_0)$, then $B(m, r_0)$ is locally convex.

When all such b_r are positive definite we call $B(m, r_0)$ *strongly locally convex*, abbreviated SLC.

Problem 28. Let $B(m, r_0)$ be a normal coordinate ball, τ a geodesic from m, and $z = \tau_*(r)$, where $r \in (0, r_0)$. Then the values of m-Jacobi fields $\{V(r) \mid V \in \mathcal{K}\}$ form the tangent space to $S(m, r)$ at $\tau(r)$. Show that:

(a) $\mathcal{K}$ is the space of $S(m, r)$-Jacobi fields along τ, so m is a focal point of $S(m, r)$ of order $d - 1$.

(b) The second fundamental form H_z of $S(m, r)$ is essentially the same as b_r.

Problem 29. Show by continuity considerations that $B(m, r)$ is SLC for sufficiently small r. If $B(m, r)$ is locally convex but not SLC, then $B(m, r_1)$ is not SLC for $r_1 > r$. Find examples to show that $B(m, r)$ can be normal for all r, SLC for $r \in (0, a)$, locally convex for $r \in [a, b]$, and not locally convex for $r > b$, where a, b are arbitrary except for the restriction $a \leqslant b \leqslant \infty$.

Problem 30. A *minimal* submanifold is one for which the second fundamental forms all have trace zero, that is, for every normal z, tr $S_z = 0$. Show that a compact minimal submanifold cannot be immersed in a strongly locally convex ball.

Problem 31. Let $B(m, r_0)$ be a strongly locally convex ball and let N be a compact manifold which is immersed in $B(m, r_0)$, giving N the induced metric. Suppose $1 + 2 \dim N > d = \dim M$. Use the following theorem of Otsuki [68, 69] to show that there is a plane section P such that $K_N(P) > K_M(P)$.

Let V be a real vector space of dimension n. Suppose $Q_1, ..., Q_k$, $k < n$, are symmetric bilinear real-valued forms on V such that

$$\sum_j (Q_j(u, u) \, Q_j(v, v) - Q_j(u, v)^2) \leqslant 0$$

for all $u, v \in V$.

Then there is at least one nonzero vector $u \in V$ such that $Q_j(u, u) = 0$ for all j.

Prove this theorem in the case $k = 1$.

Problem 32. Show that every ball in Euclidean d-space is strongly locally convex. Then apply the preceding problem to show that a flat

e-dimensional torus cannot be imbedded isometrically in E^d unless $d \geqslant 2e$.

To get more specific results on the size of SLC balls the following theorem can be used.

Theorem 13. Let H be a positive upper bound on sectional curvatures of M and let $r_0 = \pi/2H^{1/2}$. Then every normal coordinate ball having radius $r \leqslant r_0$ is SLC. If sectional curvatures are nonpositive, then every normal coordinate ball is SLC.

Proof. We have for every m-Jacobi field V, $I = I(m, N)$ as above,

$$b_r(V) = I(V) = \int_0^r (\langle V', V' \rangle - K(\tau_*, V)\langle V, V \rangle)\, du$$

$$\geqslant \int_0^r (\langle V', V' \rangle - H \langle V, V \rangle)\, du.$$

This last expression looks like the second variation of a vector V on a Riemannian manifold P which has constant curvature H. It is not difficult to see that it may be considered as such, and hence we may apply the basic inequality provided that $r < \pi/H^{1/2} = $ the distance to the first conjugate points on P. The m-Jacobi fields on P have the form $W = \sin (H^{1/2}u)\, E$, where E is parallel. Letting $E(r) = V(r)$, the basic inequality on P gives

$$b_r(V) \geqslant \langle W'(r), W(r) \rangle$$

$$= H^{1/2} \sin (H^{1/2}r) \cos (H^{1/2}r) \langle E, E \rangle.$$

Thus b_r is positive definite if $r < r_0$, as desired.

Problem 33. Let C be a compact subset of M. Show that there is $r > 0$ such that for every $m \in C$, $B(m, r)$ is convex and SLC.

Problem 34. If M is complete, simply connected, and has nonpositive curvature, show that every ball in M is convex.

Problem 35. Let M have nonpositive curvature and let B be a convex ball containing a geodesic triangle with lengths of sides a, b, c and opposite angles α, β, γ. Show that

$$a^2 + b^2 - 2ab \cos \gamma \leqslant c^2$$

$$\alpha + \beta + \gamma \leqslant \pi \qquad [33, \text{p. } 73]$$

A generalization of problem 35 for arbitrary curvature is due to Toponogov (see [9] and [90]).

Remark. The argument employed above involving the index forms for M and P is one step in the proof of Rauch's comparison theorem. The theorey of the index form may be broadened to include the integral encountered without reference to P. This is done by replacing the *Ricci transformation* $R_X : V \to R_{XV}X$ by an arbitrary smooth field S of symmetric linear transformations of the normal spaces to τ. This is general enough for our purposes, but Morse considered even more general forms [57].

If N and P are end manifolds we retain the end terms, so we get a quadratic form I_S given by

$$I_S(V) = \langle S_{\tau_*(0)} V(0), V(0) \rangle - \langle S_{\tau_*(b)} V(b), V(b) \rangle$$
$$+ \int_0^b (\langle V', V' \rangle - \langle SV, V \rangle)\, du.$$

An *N-S-field* V is a field along τ which satisfies the end conditions at N and the equation $V'' + SV = 0$. An S-focal point of N is a point on τ where a nonzero N-S-field vanishes. The basic inequality is valid and takes the form:

Suppose there are no S-focal points of N on τ. For $V \in \mathscr{L}$ there is a unique N-S-field Y such that $Y(b) = V(b)$. Then $I_S(V) \geqslant I_S(Y)$ and equality holds only if $V = Y$.

The most important application will be the case where S is derived from the Ricci transformation of some Riemannian manifold. The above proof, where $S =$ identity, is typical (cf. problem 18).

11.9 Rauch's Comparison Theorem [9, 74, 75]

We have already considered a special case, theorem 9.2, of Rauch's theorem, and some of the same ideas are involved in the proof of theorem 13. In the case $d = 2$, the analytic content of Rauch's theorem is essentially the same as Sturm's comparison theorem for second order ordinary differential equations.

Theorem 14. Let M and N be Riemannian manifolds, σ and τ geodesic segments parametrized by arc length on $[0, b]$ and starting at $m \in M$ and $n \in N$, respectively. Suppose that there is no conjugate point of n on τ. Finally, assume that for all plane sections P and Q

tangent to σ and τ at $\sigma(r)$ and $\tau(r)$, respectively, for all $r \in [0, b]$, $K_M(P) \leqslant K_N(Q)$.

Let $x = b\sigma_*(0)$, $y = b\tau_*(0)$, $s \in (M_m)_x$, $t \in (N_n)_y$. Then if s and t have the same length

$$\| d \exp_m s \| \geqslant \| d \exp_n t \|.$$

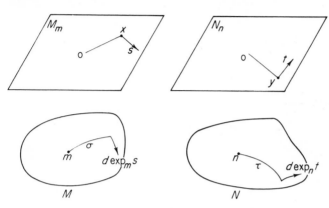

Fig. 47

Proof. Let V be the m-Jacobi field such that $V(b) = d \exp_m s$, W the n-Jacobi field such that $W(b) = d \exp_n t$, $f = \langle V, V \rangle$, and $g = \langle W, W \rangle$. We wish to show that $f(b) \geqslant g(b)$. It will be sufficient to prove

(a) $(f/g)(0_+) = 1$,

(b) $f'/f \geqslant g'/g$ on $(0, b]$, because then f/g will be nondecreasing.

To prove (a) we note that there are constant vector fields A and B on the rays to x and y in M_m and N_n such that $A(x) = s$, $B(y) = t$, $V = uX$, and $W = uY$, where $X = d \exp_m A$, $Y = d \exp_m B$. Then $f/g = \langle X, X \rangle / \langle Y, Y \rangle$. But $\langle X, X \rangle(0) = \langle A, A \rangle = \langle s, s \rangle = \langle t, t \rangle = \langle Y, Y \rangle(0)$.

To prove (b), suppose first that there is no conjugate point of m on $\sigma((0, r])$. Then $f(r) \neq 0$, so we may let $X = V/f(r)^{1/2}$; similarly, let $Y = W/g(r)^{1/2}$. Then X and Y are Jacobi fields, so

$$(f'/f)(r) = \langle X, X \rangle'(r)$$

$$= 2 \int_0^r (\langle X', X' \rangle - K_M(\sigma_*, X) \langle X, X \rangle) \, du$$

$$\geqslant 2 \int_0^r (\langle X', X' \rangle - K_N(\tau_*, X_N) \langle X, X \rangle) \, du,$$

where X_N is any nonzero vector field normal to τ. By choosing parallel bases E_i and F_i along σ and τ such that $X(r) = E_1(r)$, $Y(r) = F_1(r)$, and using the coefficients of X with respect to E_i as the coefficients of X_N with respect to F_i, we get an X_N which coincides with Y at r and for which $\langle X', X' \rangle = \langle X_N', X_N' \rangle$ and $\langle X, X \rangle = \langle X_N, X_N \rangle$. By the basic inequality for the index form along τ we then have

$$(f'/f)(r) \geqslant 2 \int_0^r (\langle Y', Y' \rangle - K_N(\tau_*, Y)\langle Y, Y \rangle)\, du$$

$$= \langle Y, Y \rangle'(r)$$

$$= (g'/g)(r).$$

It now follows that $f(r) \geqslant g(r)$ as long as there is no conjugate point of m on $\sigma((0, r])$. However, this was only needed to enable us to divide by $f(r)^{1/2}$, so the inequality and continuity now give that $f(r) \geqslant g(r)$ for $r \in (0, b]$.

Corollary 1.   Under the above hypothesis, the first conjugate point of n must occur before that of m.

Corollary 2.   (Bonnet).   Let M have all curvatures of plane sections tangent to a geodesic γ starting at m satisfy the inequalities $0 < L \leqslant K(P) \leqslant H$, L and H constants. Then if s is the distance along γ to the first conjugate point of m on γ,

$$\pi/H^{1/2} \leqslant s \leqslant \pi/L^{1/2}.$$

Proof.   This follows immediately from corollary 1 and by comparison with spheres of constant curvature L and H.

Remark.   Another consequence is that the hypothesis of Synge's theorem, theorem 11, may be weakened to completeness and strictly positive curvature instead of compactness.

Problem 36.   Generalize Rauch's theorem by replacing the points m and n by totally geodesic submanifolds of equal dimensions.

Rauch's comparison theorem and Klingenberg's results on closed geodesics are among techniques used in the study of "pinched" manifolds as well as for the proof of Topogonov's theorem on geodesic triangles [6-9, 45-47, 74-76, 89-91].

11.10 Curvature and Volume [*10, 11, 32*]

The *Ricci transformations* $R_x : y \to R_{xy}x$ extend to derivations of the Grassmann algebra; since R_x is symmetric with respect to $\langle \, , \, \rangle$, the extensions are symmetric with respect to the natural extension of $\langle \, , \, \rangle$ (problem 4.14). Let $y_1, ..., y_p$, x be orthonormal vectors and P the p-plane spanned by $y_1, ..., y_p$. The p-*mean curvature* of x and P is the inner product $K(x, P) = \langle R_x(y_1 \cdots y_p), y_1 \cdots y_p \rangle$. In particular, there is a unique $d - 1$-plane orthogonal to x, and its $d - 1$ mean curvature is called simply the *mean* (or *Ricci*) curvature of x, $K(x, x^\perp)$.

Problem 37. Show that $K(x, P) = \Sigma K(x, y_i)$, and hence that such sums depend only on the plane the y_i span. Moreover, $K(x, x^\perp) = \operatorname{tr} R_x$.

Let γ be a geodesic starting at m, $\gamma = \exp_m \circ \, \rho$, where ρ is a ray in M_m, parametrized by arc length. Let $J_p(t)$ be the maximum of all the factors by which $\exp_m$ multiplies lengths of decomposable p-vectors normal to ρ at $\rho(t)$, that is, $J_p(t)$ is the maximum of the ratios

$$\| d \exp_m s_1 \cdots d \exp_m s_p \|/\| s_1 \cdots s_p \|,$$

where s_i are linearly independent tangents to M_m normal to ρ at $\rho(t)$. Similarly, let $j_p(t)$ be the minimum of such ratios. In particular, $J_{d-1}(t) = j_{d-1}(t)$ is the Jacobian determinant of $\exp_m$ at $\rho(t)$. Note that $J_p(0) = j_p(0) = 1$.

Theorem 15. Suppose m has no conjugate point on $\gamma((0, c])$. Let (p) be the condition: For every t and every p-plane P normal to γ at $\gamma(t)$, $K(\gamma_*(t), P) \geqslant pa^2$.

(1) If (p) then $J_p(t) \leqslant (\sin at/at)^p$ for $t \in (0, c]$.

(2) Suppose that for every t and every vector y normal to γ at $\gamma(t)$, $K(\gamma_*(t), y) \leqslant b^2$. Then $j_p(t) \geqslant (\sin bt/bt)^p$ for $t \in (0, c]$.

In case $p = d - 1$ we can assert more, namely, in (1) that $J_{d-1}(t)(at/\sin at)^{d-1}$ is a nonincreasing function of t, and in (2) that $J_{d-1}(t)(bt/\sin bt)^{d-1}$ is a nondecreasing function of t.

We do not assume that a, b are real, but use the complex extension of sin in case a^2 or b^2 is negative; if a or $b = 0$ we replace $at/\sin at$ or $bt/\sin bt$ by 1.

Proof. Let $s_1, ..., s_p$ be independent vectors normal to ρ at $\rho(t)$.

The s_i generate linear homogeneous fields along ρ, which project under $d\exp_m$ to p independent m-Jacobi fields. Thus we have a smooth field of p-planes, P, spanned by these m-Jacobi fields at the points along γ. Let $f : [0, t] \to R$ be the function which gives the ratios corresponding to P, so $f(0) = 1$, and $f(t)$ is a typical ratio of which $J_p(t)$ is the maximum and $j_p(t)$ the minimum. In particular, if $p = d - 1$ then P is unique and $f = J_{d-1} = j_{d-1}$.

If $Y_1, ..., Y_p$ are m-Jacobi fields which span P at one value in $(0, t]$, then they will span P at every value. Letting $Y_i = d\exp_m uA_i$, where A_i are constant fields along ρ in M_m, then $f = \| Y_1 \cdots Y_p \|/u^p A$, where $A = \| A_1 \cdots A_p \|$ is constant. Now suppose $Y_1(r), ..., Y_p(r)$ are orthonormal; for each $r \in (0, t]$ there is such a set of Y_i. Then

$$\langle Y_1 \cdots Y_p, Y_1 \cdots Y_p \rangle'(r)$$

$$= 2 \sum \langle Y_1 \cdots Y_i' \cdots Y_p, Y_1 \cdots Y_p \rangle(r)$$

$$= 2 \sum \langle Y_i', Y_i \rangle(r),$$

which follows from the fact that the basis of p-vectors generated by an orthonormal basis of vectors is orthonormal. [Express $Y_i'(r)$ in terms of an orthonormal basis which includes $Y_1(r), ..., Y_p(r)$ and expand.] Using this to differentiate f^2 yields the relation

(a) $$f'(r)/f(r) = \sum \langle Y_i', Y_i \rangle(r) - p/r.$$

By the basic inequality, if W_i is a vector field along γ such that $W_i(0) = 0$ and $W_i(r) = Y_i(r)$, then

$$\langle Y_i', Y_i \rangle(r) = I_r(Y_i) \leqslant I_r(W_i),$$

where $I_r = I(m, N_r)$ with intermediate submanifold N_r having null second fundamental form. In particular, if E_i is the parallel field generated by $Y_i(r)$ and g is a broken C^∞ function such that $g(0) = 0$, $g(r) = 1$, taking $W_i = gE_i$ gives

$$\langle Y_i', Y_i \rangle (r) \leqslant \int_0^r ((g')^2 - K(\gamma_*, E_i) g^2)\, du.$$

Adding these inequalities and using the hypothesis

$$(p)\colon \sum K(\gamma_*, E_i) \geqslant pa^2,$$

it follows from (a) that

$$f'(r)/f(r) \leqslant p \left(\int_0^r ((g')^2 - a^2 g^2) \, du - 1/r \right).$$

The integral in this inequality is the second variation of a vector field gE, where E is parallel and unitary, in a space of constant curvature a^2. It follows that the best choice for g is that which will make gE a Jacobi field, by the basic inequality. Thus we let $g = \sin au/\sin ar$, and the integral becomes

$$\langle g'E, gE \rangle(r) = a \cos ar/\sin ar.$$

(If $a = 0$ let $g = u/r$.) Now we integrate the resulting inequality:

(b) $$f'(r)/f(r) \leqslant p(a \cos ar/\sin ar - 1/r)$$

from q to t, $q \in (0, t)$ and take exponentials, obtaining

$$f(t)(at/\sin at)^p \leqslant f(q)(aq/\sin aq)^p.$$

In case $p = d - 1$ this yields the desired monotonicity of $J_{d-1}(t) (at/\sin at)^{d-1}$. Otherwise we take the limit as $q \to 0_+$, using the fact that $f(0) = 1$, so

$$f(t) \leqslant (\sin at/at)^p.$$

This is true for all such $f(t)$, hence also for their maximum $J_p(t)$. This completes the proof of (1).

To prove (2) we return to the inequality (a) and use the hypothesis $K(\gamma_*, Y_i) \leqslant b^2$:

$$\langle Y_i', Y_i \rangle(r) = \int_0^r (\langle Y_i', Y_i' \rangle - K(\gamma_*, Y_i) \langle Y_i, Y_i \rangle) \, du$$

$$\geqslant \int_0^r (\langle Y_i', Y_i' \rangle - b^2 \langle Y_i, Y_i \rangle) \, du$$

$$\geqslant \langle h'E_i, hE_i \rangle(r),$$

where hE_i is a "Jacobi field" for a space of constant curvature b^2, the last step following from the basic inequality; thus $h = \sin bu/\sin br$. Now we proceed as before, obtaining (b) with the inequality reversed and a replaced by b, and so on to the conclusion of (2).

Corollary 1. If (p) and $a^2 > 0$, then the first conjugate point of m on γ occurs at least within distance π/a along γ.

[$\rho(t)$ is a conjugate point of m if and only if a nonzero p-vector is annihilated by $d \exp_m \cdot$]

Corollary 2. (Myers' theorem [62]). If M is a complete Riemannian manifold with mean curvature bounded away from 0 by $(d - 1)a^2 > 0$, then M is compact, the diameter of M is $\leqslant \pi/a$, and the fundamental group of M is finite.

Proof.   The first two assertions follow from corollary 1. The simply connected Riemannian covering of M has the same local properties, so must also be compact. But the fibres of a covering space are discrete, so in a compact covering they are finite, and hence the fundamental group is finite.

Corollary 3.   Let $v(m, r)$, for r sufficiently small, denote the volume of the sphere $S(m, r)$ contained in a normal coordinate neighborhood. If $(d - 1) a^2$ is a lower bound for mean curvature on M, and b^2 is an upper bound for curvature, then $v(m, r) (a/\sin ar)^{d-1}$ is a non-increasing function of r, $v(m, r) (b/\sin br)^{d-1}$ is a nondecreasing function of r.

Proof.   Since we want to consider all geodesics from m, we let $J(r, x)$ denote $J_{d-1}(r)$ from above when $x = \gamma_*(0)$. $J(r, x)$ is the Jacobian of the restriction of $\exp_m$ to the sphere S_r of radius r in M_m. Combining $\exp_m$ with the map $r \to rx$ we have a map of S_1 onto $S(m, r)$ with Jacobian $r^{d-1} J(r, x)$. Thus

$$v(m, r) = \int_{S_1} r^{d-1} J(r, x)\, dx,$$

so the result now follows from the fact that $J(r, x) (ar/\sin ar)^{d-1}$ is monotonic in r, and similarly for the other case with b.

Remark.   The same conclusion obtains if curvature has range $[a^2, b^2]$, which is a result of Günther [32].

Corollary 4.   If M is complete and $(d - 1) a^2$ is a lower bound for mean curvature, then the volume of a normal coordinate ball $B(m, r)$ is $\leqslant$ the volume of a normal coordinate ball of the same size in the *simple space form* (that is, sphere, Euclidean space, or hyperbolic space) of constant curvature a^2.

If $a^2 > 0$ the volume of M is $\leqslant$ the volume of a sphere of radius $1/a$, and equality obtains only if M is isometric to such a sphere.

Proof. The volume of $B(m, r)$ is obtained by integrating the Jacobian of $\exp_m$ on a ball of radius r in M_m. But the bound $(\sin ar/ar)^{d-1}$ is the Jacobian of an exponential map of the space form. (This follows from the proof of theorem 15, since constant curvature gives equality throughout in that proof.)

If $a^2 > 0$, then M is compact and the volume is given by integrating the Jacobian of $\exp_m$ on the open set of M_m which is within the minimum locus. The integral of the bound for that Jacobian, $(\sin ar/ar)^{d-1}$ on as great an open set, the ball of radius π/a, will be no less, but gives the volume of a sphere of radius $1/a$.

If the volume of M equals that of a sphere of radius $1/a$, then all of the inequalities in the proof of theorem 15(1) must be equations; in particular, the Jacobi fields of M are in the same form as those on a sphere of radius $1/a$, so M has constant curvature a^2 and is locally isometric to a sphere of radius $1/a$. But the minimum locus in M_m must be the sphere S of radius π/a, and since $d \exp_m$ annihilates all vectors tangent to S, $\exp_m(S)$ is a point. Thus by factoring through the exponential maps a global isometry is obtained.

Problem 38. Show that the conditions (p) are monotone in strength, that is, (p) implies $(p + 1)$ (with the same a).

Problem 39. Let S be the simple space form of curvature a^2, M a manifold satisfying (p) along all geodesics from m, N either

(1) a p-dimensional submanifold of M_m such that it is contained in a sphere with center O and within a normal coordinate ball, or

(2) a $(p + 1)$-dimensional submanifold of M_m within a normal coordinate ball.

Let $\phi : M_m \to S_{m'}$ be a linear isometry. Show that

$$\text{volume } (\exp_m(N)) \leqslant \text{volume } (\exp_{m'} \circ \phi\, (N)).$$

[*Hint*: For (2) an extension of theorem 15 to arbitrary $(p + 1)$-vectors is needed. Use the fact that if $x = \gamma_*(r)$, an arbitrary decomposible $(p + 1)$-vector may be written $(x + y_1) y_2 \cdots y_{p+1}$, where y_i are all normal to x.]

Appendix. Theorems on Differential Equations

The reader may find it instructive to translate (i) and (ii) below into their coordinate forms and verify that essentially the same theorems are found in standard references [26].

Let F be a C^∞ map, $F : U^n \times U^m \to T(R^n)$, where U^n and U^m are open subsets of R^n and R^m, respectively, $T(R^n)$ the tangent bundle to R^n, and such that $F(u, u') \in R_u{}^n$ for every $(u, u') \in U^n \times U^m$. Thus F gives a C^∞ vector field on U^n for every $u' \in U^m$; or, in classical terminology, a system of n first order differential equations in n unknowns, depending in addition on a parameter $u' \in U^m$.

(i) Existence and Uniqueness

There is a unique C^∞ map $\phi : W \to R^n$, where W is a neighborhood of $\{0\} \times U^n \times U^m$ in $R \times U^n \times U^m$, such that for every $p = (t, u, u') \in W$:

(a) $d\phi_p(D_1(p)) = F(\phi(p), u')$, where D_1 is the partial derivative operator corresponding to R and its coordinate u_1 in $R \times U^n \times U^m$.

(b) $\phi(0, u, u') = u$.

(ii) Continuation of Solutions

Suppose that F is bounded in the Euclidean metric on R^n. Then the neighborhood W above can be taken so that its intersection with the fibres of the projection $R \times U^n \times U^m \to U^n \times U^m$ are of the form $(a, b) \times \{u\} \times \{u'\}$ and either $b = \infty$ or $\lim_{t \to b-} \phi(t, u, u')$ exists and is outside U^n, and similarly for the other end. a and b depend on u and u'.

This says that integral curves can be continued in either direction until either the parameter becomes infinite or the curve runs out of U^n.

(iii) Extension to Manifolds

The above results are true if U^n is an open subset of a manifold N,

U^m an open subset of a manifold M,

$$F : U^n \times U^m \to T(N).$$

(iv) The Local Group Associated with F, $u' \in U^m$

For each $u' \in U^m$, t, $s \in R$ such that the following are defined:

(a) $\phi \mid_{\{t\} \times U^n \times \{u'\}}$ is a diffeomorphism.

(b) $\phi(s, \phi(t, u, u'), u') = \phi(s + t, u, u')$.

Bibliography

1. Ambrose, W. The index theorem in Riemannian geometry. *Ann. of Math.* **73** (1961), 49-86.
2. Ambrose, W., and Singer, I. M. A theorem on holonomy. *Trans. Amer. Math. Soc.* **75** (1953), 428-443.
3. Atiyah, M. Complex analytic connections in fibre bundles. *Trans. Amer. Math. Soc.* **85** (1957), 181-207.
4. Auslander, L., and MacKenzie, R. E. "Introduction to Differentiable Manifolds." McGraw-Hill, New York, 1963.
5. Berger, M. Sur les groupes d'holonomie homogène des variétés à connexion affine et des variétés Riemanniennes. *Bull. Soc. Math. France* **83** (1955), 279-330.
6. Berger, M. Les variétés riemanniennes dont la courbure satisfait à certaines conditions. *Proc. Internat. Congr. of Mathematicians Stockholm, 1962.* 447-456.
7. Berger, M. Pincement riemannien et pincement holomorphe. *Ann. Scuola Norm. Sup. Pisa* [Ser. III], **14** (1960), 151-159.
8. Berger, M. Les variétés riemanniennes (1/4)-pincées. *Ann. Scuola Norm. Sup. Pisa* [Ser. III], **14** (1960), 161-170.
9. Berger, M. An extension of Rauch's metric comparison theorem and some applications. *Illinois J. Math.* **6** (1962), 700-712.
10. Bishop, R. A relation between volume, mean curvature, and diameter. *Amer. Math. Soc. Not.* **10** (1963), 364.
11. Bishop, R., and Goldberg, S. Some implications of the generalized Gauss-Bonnet Theorem. *Trans. Amer. Math. Soc.* To appear.
12. Bochner, S., and Yano, K. "Curvature and Betti Numbers," Ann. Math. Studies No. 32. Princeton Univ. Press, Princeton, New Jersey, 1953.
13. Borel, A. Lectures on symmetric spaces. M.I.T. lecture notes (1958).
14. Bott, R., and Milnor, J. On the parallelizability of the spheres. *Bull. Amer. Math. Soc.* **64** (1958), 87-89.
15. Cartan, É. Sur l'intégration des systèmes d'équations aux différentielles totales. *Ann. Sci. École Norm. Sup.* **18** (1901), 241-311.
16. Cartan, É. Sur la possibilité de plonger un espace riemannien donné dans un espace euclidien. *Ann. Soc. Polon. Math.* **6** (1927), 1-7.
17. Cartan, É. "Leçons sur la géométrie des espaces de Riemann." Gauthier-Villars, Paris, 1946.
18. Cartan, É. Sur une classe remarquable d'espaces de Riemann, I; II. *Bull. Soc. Math. France* **54** (1926), 214-264; **55** (1927), 114-134.
19. Chern, S.-S. A simple intuitive proof of the Gauss-Bonnet formula for closed Riemannian manifolds. *Ann. of Math.* **45** (1944), 747-752.
20. Chern, S.-S. Topics in differential geometry. Inst. Advanced Study lecture notes (1951).

21. Chern, S.-S. An elementary proof of the existence of isothermal parameters on a surface. *Proc. Amer. Math. Soc.* **6** (1955), 771-782.

22. Chern, S.-S. On curvature and characteristic classes of a Riemann manifold. *Abh. Hamburger Univ. Math. Sem.* **20-21** (1955), 117-126.

23. Chern, S.-S. Complex manifolds. Univ. of Chicago lecture notes (1956).

24. Chern, S.-S. Differentiable manifolds. Univ. of Chicago lecture notes (1959).

25. Chevalley, C. "Theory of Lie Groups." Princeton Univ. Press, Princeton, New Jersey, 1946 (2nd printing, 1960).

26. Coddington, E., and Levinson, N. "Theory of Ordinary Differential Equations." McGraw-Hill, New York, 1955.

27. Cohn, P. M. "Lie Groups." Cambridge Univ. Press, London and New York, 1957.

28. Crittenden, R. Minimum and conjugate points in symmetric spaces. *Canad. J. Math.* **14** (1962), 320-328.

29. Flanders, H. "Differential forms", Academic Press, New York, 1963.

30. Goldberg, S. I. "Curvature and Homology." Academic Press, New York, 1962.

31. Grauert, H. On Levi's problem and the imbedding of real-analytic manifolds. *Ann. of Math.* **68** (1958), 460-472.

32. Günther, P. Einige Sätze über das Volumenelement eines Riemannschen Raumes. *Publ. Math. Debrecen* **7** (1960), 78-93.

33. Helgason, S. "Differential Geometry and Symmetric Spaces." Academic Press, New York, 1962.

34. Hellwig, G. "Partielle Differentialgleichungen." Teubner, Stuttgart, 1960.

35. Hermann, R. Cartan connexions and the equivalence problem for geometric structures. Univ. of California (Berkeley), N.S.F. report (1962).

36. Hermann, R. É. Cartan's theory of exterior differential systems. Univ. of California (Berkeley) N.S.F. report (1963).

37. Hermann, R. Focal points of closed submanifolds of Riemannian spaces. *Nederl. Akad. Wetensch. Proc. Ser. A* **66** (1963), 613-628.

38. Hermann, R. "Dynamical Systems and the Calculus of Variations." Academic Press, New York, in preparation.

39. Hocking, J. G., and Young, G. S. "Topology." Addison-Wesley, Reading, Massachusetts, 1961.

40. Hopf, H., and Rinow, W. Über den Begriff der vollständigen differentialgeometrischen Fläche. *Comment. Math. Helv.* **3** (1931), 209-225.

41. Hu, S. T. "Homotopy Theory." Academic Press, New York, 1959.

42. Jacobson, N. "Lie Algebras." Wiley (Interscience), New York, 1962.

43. Kähler, E. Einführung in die Theorie der Systeme von Differentialgleichungen. *Hamburger Math. Einzelschr.* **16** (1934).

44. Kervaire, M. A manifold which does not admit any differentiable structure. *Comment. Math. Helv.* **35** (1961), 1-14.

45. Klingenberg, W. Contributions to Riemannian Geometry in the large. *Ann. of Math.* **69** (1959), 654-666.

46. Klingenberg, W. Über kompakte Riemannsche Mannigfaltigkeiten. *Math. Ann.* **137** (1959), 351-361.

47. Klingenberg, W. Über Riemannsche Mannigfaltigkeiten mit positiver Krümmung. *Comment. Math. Helv.* **35** (1961), 47-54.

48. Kobayashi, S. On connections of Cartan. *Canad. J. Math.* **8** (1956), 145-156.

49. Kobayashi, S. Theory of connections. *Ann. Mat. Pura Appl.* **43** (1957), 119-194.

50. Kobayashi, S., and Nomizu, K. "Foundations of Differential Geometry." Wiley (Interscience), New York, Vol. I, 1963; Vol. II, in preparation.
51. Lichnerowicz, A. "Théorie globale des connexions et des groupes d'holonomie." Edizione Cremonese, Rome, 1955.
52. Milnor, J. On manifolds homeomorphic to the 7-sphere. *Ann. of Math.* **64** (1956), 399-405.
53. Milnor, J. Differential topology. Princeton Univ. lecture notes (1958).
54. Milnor, J. "Lectures on Morse Theory," Ann. Math. Studies No. 51, Princeton Univ. Press, Princeton, New Jersey, 1963.
55. Montgomery, D., and Zippin, L. "Topological Transformation Groups." Wiley (Interscience), New York, 1955.
56. Morrey, C. B. The analytic embedding of abstract real-analytic manifolds. *Ann. of Math.* **68** (1958), 159-201.
57. Morse, M. "The Calculus of Variations in the Large." Amer. Math. Soc., New York, 1934.
58. Munkres, J. R. "Elementary Differential Topology," Ann. Math. Studies No. 54, Princeton Univ. Press, Princeton, New Jersey, 1963.
59. Myers, S. Riemannian manifolds in the large. *Duke Math. J.* **1** (1935), 39-49.
60. Myers, S. Connections between differential geometry and topology, I. Simply connected surfaces. *Duke Math. J.* **1** (1935), 376-391.
61. Myers, S. Connections between differential geometry and topology, II. Closed surfaces. *Duke Math. J.* **2** (1936), 95-102.
62. Myers, S. Riemannian manifolds with positive mean curvature. *Duke Math. J.* **8** (1941), 401-404.
63. Nash, J. The imbedding problem for Riemannian manifolds. *Ann. of Math.* **63** (1956), 20-63.
64. Newlander, A., and Nirenberg, L. Complex analytic coordinates in almost complex manifolds. *Ann. of Math.* **65** (1957), 391-404.
64A. Newns, W.F. and Walker, A.G., Tangent planes to a differentiable manifold, *J. London Math. Soc.* **31** (1956), 400-407.
65. Nomizu, K. Invariant affine connections on homogeneous spaces. *Amer. J. Math.* **76** (1954), 33-65.
66. Nomizu, K. "Lie Groups and Differential Geometry." Math. Soc. of Japan, Tokyo, 1956.
67. Nomizu, K., and Ozeki, H. The existence of complete Riemannian metrics. *Proc. Amer. Math. Soc.* **12** (1961), 889-891.
68. Otsuki, T. On the existence of solutions of a system of quadratic equations and its geometric applications. *Proc. Japan Acad.* **29** (1953), 99-100.
69. Otsuki, T. Isometric imbedding of Riemann manifolds in a Riemann manifold. *J. Math. Soc. Japan* **6** (1954), 221-234.
70. Ozeki, H. Infinitesimal holonomy groups of bundle connections. *Nagoya Math. J.* **10** (1956), 105-123.
71. Pitcher, E. Inequalities of critical point theory. *Bull. Amer. Math. Soc.* **64** (1958), 1-30.
72. Pontrjagin, L. "Topological Groups." Princeton Univ. Press, Princeton, New Jersey, 1946.
73. Preissmann, A. Quelques propriétés globales des espaces de Riemann. *Comment. Math. Helv.* **15** (1942-3), 175-216.

74. Rauch, H. A contribution to differential geometry in the large. *Ann. of Math.* **54** (1951), 38-55.

75. Rauch, H. "Geodesics and Curvature in Differential Geometry in the Large." Yeshiva Univ. Press, New York, 1959.

76. Rauch, H. The global study of geodesics in symmetric and nearly symmetric Riemannian manifolds. *Comment. Math. Helv.* **35** (1961), 111-125.

77. de Rham, G. Sur la réductibilité d'un espace de Riemann. *Comment. Math. Helv.* **26** (1952), 328-344.

78. de Rham, G. "Variétés différentiables." Hermann, Paris, 1955.

79. Sard, A. The measure of the critical values of differentiable maps. *Bull. Amer. Math. Soc.* **48** (1942), 883-890.

80. Seifert, H., and Threlfall, W. "Lehrbuch der Topologie." Teubner, Stuttgart, 1934.

81. Seifert, H., and Threlfall, W. "Variationsrechnung im Grossen." Chelsea, New York, 1951.

82. Simons, J. On the transitivity of holonomy systems. *Ann. of Math.* **76** (1962), 213-234.

83. Singer, I. Differential geometry. M.I.T. lecture notes (1962).

84. Smale, S. A survey of some recent developments in differential topology. *Bull. Amer. Math. Soc.* **69** (1963), 131-145.

85. Steenrod, N. "The Topology of Fibre Bundles." Princeton Univ. Press, Princeton, New Jersey, 1951.

86. Synge, J. L. The first and second variations of the length in Riemannian space. *Proc. London Math. Soc.* **25** (1926), 247-264.

87. Synge, J. L. On the connectivity of spaces of positive curvature. *Quart. J. Math. Oxford Ser.* (1), **7** (1936), 316-320.

88. Thom, R. Quelques propriétés globales des variétés différentiables. *Comment. Math. Helv.* **29** (1954), 17-85.

89. Tsukamoto, Y. On certain Riemannian manifolds with positive sectional curvature. *Mem. Fac. Sci. Kyushu Univ. Ser. A*, **13** (1959), 140-145.

90. Tsukamoto, Y. On a theorem of A. D. Alexandrov. *Mem. Fac. Sci. Kyushu Univ. Ser. A*, **15** (1962), 83-89.

91. Tsukamoto, Y. On Riemannian manifolds with positive curvature. *Mem. Fac. Sci. Kyushu Univ. Ser. A*, **15** (1962), 90-96.

92. Weil, A. "Variétés kählériennes." Hermann, Paris, 1958.

93. Whitehead, J. H. C. On the covering of a complete space by the geodesics through a point. *Ann. of Math.* **36** (1935), 679-704.

94. Willmore, T. "An Introduction to Differential Geometry." Oxford, Univ. Press, London and New York, 1959.

95. Yamabe, H. On an arc-wise connected subgroup of a Lie group. *Osaka Math. J.*, **2** (1950), 13-14.

Subject Index

265